# Company-Wide

# IMPLEMENTATIONS

# *of*

# ROBUST-TECHNOLOGY

# Development

by
**Kenzo Ueno**
PRESIDENT,
UENO DESIGN INSTITUTE
•
translated by
**Shih-Chung Tsai**

ASME PRESS

**Library of Congress Cataloging-in-Publication Data**

Ueno, Kenzo, 1942-
  [Gijutsu Saikouchiku.  English]
  Company-Wide Implementations of Robust-Technology Development / by
  Kenzo Ueno; translated by Shih-chung Tsai.
    p.  cm.
  Includes index.
  ISBN  0-7918-0050-4 (alk. paper)
  1. Quality control.    2. Taguchi methods (Quality control)
  3. Reliability  (Engineering)    I. Title.
  TS156.U3513  1997                             97-859
  658.5'62--dc21                                 CIP

# ABOUT THE AUTHOR

 Kenzo Ueno graduated from the Engineering School of the University of Tokyo, Japan. He joined Prince Motor Corporation (which later became Nissan Motor Corporation) in 1966. From 1966 to 1981, he worked as a design engineer in the Chassis Development Division, conducting design activities in suspension systems. He became Manager of the Chassis Design Department in 1982; Manager of the Chassis Testing Department in 1985; and Manager of the Vehicle Testing Department in 1987. In 1989, he became the General Manager of Reliability Engineering Center. Since his retirement from Nissan Motor Corporation in 1995, Mr. Ueno has been the President of Ueno Design Institute, in Kanagawa, Japan, providing a consulting service for worldwide robust engineering. As an expert on the implementations of robust engineering, he is a frequent lecturer and the author of three books and more than 20 articles on the subject. He won the Best Case Study Award of the 12th Taguchi Symposium in 1994 and the Best Thesis Award of Quality Engineering Forum of Japan in 1995.

# PUBLICATION

Coauthors Genichi Jaguchi, Ikuro Baba, Jakashi Kamoshira, Akira Sugiyinia, Kazuhiko Hara, and Hiroshi Yano, *Introduction to Quality Engineering Applications — Transformability and Technology Development*, Japanese Standards Association, Jolsyo, Japan
Publishing Date: November 20, 1992

Professional societies:
Board of Directors of the Quality Engineering Forum of Japan
Member of the Society of Mechanical Engineers of Japan
Member of the Society of Automotive Engineers of Japan

# ABOUT THE TRANSLATOR

 Dr. Shih-Chung Tsai works with General Motors Corporation, specializing in robust engineering and experimental design methodologies. He received his Ph.D. in Mechanical and Aerospace Engineering from the University of Missouri-Columbia in 1990. Dr. Tsai has published more than 10 articles on robust engineering in technical journals and magazines. Having spent more than eight years studying the Japanese language has enabled him to perform professional translations of numerous technical publications from Japanese into English, including Dr. Genichi Taguchi's landmark book, *Taguchi on Robust Technology Development*, ASME Press, 1993. He is a Certified Quality Engineer of the American Society for Quality Control and also a member of the Quality Engineering Forum of Japan.

# TABLE OF CONTENTS

Note from the Japanese-Version Editor     vii
Translator's Note     ix
Foreword     xi
Preface     xiii
Acknowledgments     xv

**1 Restructuring Technology Bases and Robust Engineering   1**
   1.1 From Design Department to Testing Department    3
   1.2 Efforts in Reliability Improvement    6
   1.3 Initiation of Robust Engineering    8
   1.4 The Reasons for Using Robust Engineering Methods    9
   1.5 Summary    11

**2 The Evolution of Reliability Engineering   13
in the Automobile Industry**
   2.1 Initiation of Reliability Engineering    15
   2.2 Motorization in the 1960s    16
   2.3 Emission Control in the 1970s    18
   2.4 Electronification in the 1980s    21
   2.5 Development of New Technologies in the 1990s    23

**3 Changing Conditions in Society and Strategies   25
for Solving Related Quality Problems**
   3.1 The Impact From Growth in New Technologies    27
   3.2 New Technologies That are Independent    30
      of Previous Experience
   3.3 Technology-Development Engineers and Quality Problems    33

**4 Quality Problems and Action Strategies   35**
   4.1 The Endless Struggle Against Noise Factors    37
   4.2 The Fast Advancement of New Technologies    37
   4.3 Science and Engineering    39
   4.4 Variation    41
   4.5 Emptying and Restructuring of Technology Bases    43
   **Case Study 1: Static Welding Strength and    45
   Laser-Welding Technology**

**5 The Concepts of Parameter Design (Robust Engineering) 61**
   5.1 Robustness Against Noise Factors    63

5.2 The Shift From Product-Oriented Development 64
to Technology-Oriented Development

5.3 Parameter Design (Robust Engineering) 66

5.4 Using the Nonlinearity Effects Between 68
Output Responses and Control Factors to Achieve
Robustness

**6 Company-Wide Implementations of Robust Engineering 75**
6.1 Background 77
6.2 Introducing Robust Engineering into the Company 78
6.3 Working on Challenging Projects 83
6.4 Consulting on Robust Engineering Projects 85
and Top-Down Popularization
6.5 Impact of Robust Engineering on the Company 87

**7 Classifications of Quality Characteristics 91**
**and Associated Quality Requirements**
7.1 Definition of Quality 93
7.2 Downstream Quality and Upstream Quality 94
7.3 Quality and Functionality 96
7.4 Classification of Quality Characteristics 96
7.5 Customer Requirements and Quality Characteristics 98
7.6 The Relationship Between Quality and Cost 99
7.7 Preventing Product Failures by Using Robust Engineering 100
**Case Study 2: Welding Strength and Life Cycles 101**

**8 The Basic Function of a Technology 111**
8.1 The Fundamentals of Quality Problems 113
8.2 Ball Joints and Friction Force 113
8.3 Machining a Perfectly Spherical Head and Cavity 115
8.4 Squeaking Automotive Brake Pads 117
**Case Study 3: Development of Machining 121**
**Technology for High-Strength Steel**

**9 Implementations of Robust-Technology Development 135**
9.1 New Technologies and Automobile Industries 137
9.2 Manufacturing Industries and Robust Technologies 139
**Case Study 4: Development of CFRP (Carbon-Fiber- 144**
**Reinforced Plastic) Injecting-Molding Technology**

**10 The Future of Robust Engineering 161**

**Index 169**

# NOTE FROM THE JAPANESE-VERSION EDITOR

Many Japanese engineers and managers believe that Japan, being a technical leader, can learn little from other countries. Yet even now Japanese manufacturing industries need to import many key technologies from other industrialized nations. This technological dependence has long been a problem. Mr. Ueno, who has been involved in industrial design and development activities for almost three decades, maintains that it is crucial for manufacturing companies to develop fundamental bases of robust technologies in order to efficiently meet global competition in the next century. Robust engineering is the most efficient tool for the development of robust technologies. In this book, Mr. Ueno discusses how to popularize and implement robust engineering throughout a manufacturing company.

Robust engineering has been widely discussed in both academic circles and in industry, and many theories have been developed and published in books, journals, and professional magazines. However, few of these really discuss how to implement robust engineering throughout the company. In other words, few people have discussed how to initialize and formulate robust-engineering projects. In fact, project initiation and problem formulation are the most important parts of any robust engineering project. This book discusses how to shift the quality paradigms of engineers from downstream "fire-fighting" to upstream development of robust technologies. The four case studies in this book illustrate the logic and project-formulation process behind the development of generic manufacturing technologies. The conceptual framework presented can also be used in the development of numerous key technologies in other engineering fields.

The development process of robust technologies illustrated in this book will no doubt reshape the product-development processes of all manufacturing industries.

*August 1993*

Tatsu Sawada
Manager of Publication Division,
Japanese Standards Association

# TRANSLATOR'S NOTE

Robust engineering is a combination of art and science. The science component involves orthogonal arrays, experimental design, S/N ratio, control-by-noise interactions, and so on. The art component involves engineers' paradigms of variation reduction and of project initiation and formulation. While some engineering schools have begun teaching the science component of robust engineering, few teach about the art. It is the intention of Mr. Ueno to discuss only the art component of robust engineering, which is more important from the viewpoint of robust technology development. In robust engineering, there is no substitute for good project initiation and formulation.

A base of robust technologies is critical for any manufacturing company to survive in this fiercely competitive global market. Robust technologies will serve as the building blocks for future products and associated manufacturing processes. Company-wide on-the-job training and robust engineering implementation are the keys to creating a base of robust technologies.

In many manufacturing companies, robust engineering is primarily used as a downstream "fire-fighting" tool, not as a tool for the upstream development of robust technologies. This book provides four real-life case studies to illustrate how to initiate and formulate robust engineering projects for the development of generic technologies.

This book is recommended for all engineers and managers at manufacturing companies everywhere, as well as for those engineering students who are preparing to deal with the variation effects of the real engineering world.

*December 1995*

Shih-Chung Tsai, Ph.D., CQE
Translator

# FOREWORD

M r. Kenzo Ueno has been a design engineer and manager in automotive industries for more than 25 years. The objective of this book is to illustrate the necessity of shifting engineers' paradigms from traditional downstream fire-fighting to upstream robust engineering. Robust engineering should be conducted through well-formulated functionality measurement for quality assurance of all technologies. In this high-tech era, robust engineering is an absolute necessity in all engineering areas, especially in mechanical and electronic engineering.

In order to develop high-quality products at low cost, engineers need to ensure the robustness of the basic functionality of the products at all development stages, such as material selection, product design, component manufacturing, final assembly, and so on.

The theme of the Eighth Taguchi Symposium, sponsored by ASI (American Supplier Institute), was: "To get quality, don't measure quality," meaning that measuring downstream quality characteristics cannot improve product quality efficiently. Only through robust engineering can engineers improve product quality in a cost- and time-efficient manner. Mr. Ueno was one of the few engineers who first understood the meaning of this theme. He then went on to introduce and popularize robust engineering methodologies throughout the Nissan Motor Corporation of Japan over the last several years.

One of his case studies, "NC Machining Technology Development—NC Machining for High-Frequency Heat Treatment," was presented at the Tenth Taguchi Symposium in 1992 and won the Best Case Study Award. This case study is illustrated in this book to show engineers how to initiate and formulate robust engineering projects for the development of generic technologies. Robust engineering should be a fundamental quality tool for all manufacturing companies.

When the Quality Engineering Forum of Japan was founded in 1992, Mr. Ueno was elected as one of the board of directors. He has been very anxious to spread robust engineering paradigms and methodologies to the manufacturing industries of Japan and other countries. He is a colleague and friend, and one of the best engineers around when it comes to developing robustness measurements for the basic functions of generic technologies.

I recommend this book not only for every engineer, but also for every manager in all manufacturing fields.

*October 1995*

Genichi Taguchi

# PREFACE

J apanese lean production systems may be said to have been a powerful tool in supporting Japanese manufacturing industries in global competition up until now. Unfortunately, these systems have reached their limits, and several major Japanese manufacturing companies are struggling to survive in the fiercely competitive global market. Global competition has forced major international manufacturing industries to restructure their organizations. As a result, the question of how to restructure an organization has become a popular topic all over the world. The objective of restructuring is to improve the productivity of white- and blue-collar workers. However, many companies limit their efforts to reducing the head count to meet benchmark productivity requirements, a practice that in itself cannot efficiently increase productivity. Manufacturing industries need to develop more efficient strategies to improve their productivity if they are to survive in the global market.

Restructuring robust-technology bases is an efficient way for manufacturing industries to increase productivity and to compete in the global market. We know that the three major markets of the world—Japan, Europe, and North America—are already oversaturated. It is an uphill battle for any manufacturing company to increase market share there. Product-oriented development processes, which are based on Japanese lean production systems, are neither time nor cost efficient enough to provide a sufficient competitive edge in the global market. An alternative is to develop flexible and robust technologies. Simply put, we need to develop robust technologies and accumulate all associated technological information so as to restructure robust-technology bases. The newly developed technologies can then be applied to develop a family of robust new products. It is absolutely necessary to develop and accumulate robust technologies to be technologically competitive.

The objective of this book is to examine the question of how to restructure robust-technology bases so as to enhance the competitiveness of manufacturing industries. In this book, I propose strategies to popularize robust engineering and to develop robust-technology bases within any given manufacturing company.

This book is written based on my nearly three decades of experience in manufacturing industries. Its key point is the thinking and project-formulating processes behind the development of robust technologies. The experimental design methods and analytical details of robust engineering are indeed not as important as the thinking and project-formulating processes; thus, I have kept the former to a minimum. I hope that readers can understand the objective of this book and apply its concepts to develop robust technologies.

*January 1996*

Kenzo Ueno
Ueno Design Institute
Japan

# ACKNOWLEDGMENTS

It is with the deepest appreciation that I learned robust engineering directly from its founder, Dr. Genichi Taguchi, through his invaluable guidance. I also thank Mr. Tatsu Sawada of the Japanese Standards Association for his management efforts to make this project possible. I am indebted to Dr. Hilario L. Oh of General Motors Corporation and Dr. Don Clausing of M.I.T. for their comments and support for the publication of this book. Thanks are also due to Mr. Shin Taguchi and Professor Yuin Wu of American Supplier Institute, Inc., for allowing me to reprint two case studies in this book. I appreciate the proofreading and editorial help from Mr. Walter R. Langlois. Finally, I am much grateful to Dr. Shih-Chung Tsai for translating the Japanese original of this book into English.

## ACKNOWLEDGMENTS

# Restructuring Technology Bases and Robust Engineering

## 1.1  FROM DESIGN DEPARTMENT
## TO TESTING DEPARTMENT

The author joined Nissan Motor Corporation 27 years ago and spent 19 years in the chassis-development division, focusing primarily on suspension-system design. Although engineers in both the chassis design and development departments rigorously pursued development, failures caused by reasons both explainable and unexplainable invariably occurred in field tests and even later, as when cars were used by customers. Engineers felt that they never had enough resources or time to handle all these downstream problems.

After being transferred to the chassis-testing department, the author found that the working atmosphere in the testing department was different from that of the design department. The engineers of the design department were quite creative and proactive in developing new products, while those of the test department focused their resources on routine tests. This wide gulf in working attitudes and paradigms between engineers in the design and testing departments encouraged the author to consider how the company could develop high-quality automobiles smoothly and productively.

The testing department seemed to have much less technological information than the design department, which was about one kilometer distant. This one kilometer seemed to constitute a tremendous barrier between the two departments, a condition that did not escape the notice of most engineers transferred to the testing department. The testing department, located in the technical center, had more technological information than either the company's Murayama or Tochigi Proving Grounds. Similarly, there were big differences between the technical center in Atsugi, and the two proving grounds. Strangely, personnel at the testing department and the two proving grounds never expressed this opinion.

The author felt that in terms of task assignments, the testing department was somewhat like a subcontractor for the design department. The author was skeptical about testing standards and procedures used at the testing department and the two proving grounds because the tests that had been established years earlier were based on some unclear assumptions and information, and had never been upgraded. The author also was doubtful of the applicability of these testing standards and procedures because many social factors, such as road conditions, traffic regulations, emission regulations, etc., had changed over the years.

For example, the percentage of paved roads in Japan has increased dramatically since the 1960s, resulting in significant improvements in the high-speed performance of automobiles. Improvement in road conditions has also reduced the vertical loads on automobile suspension systems. However, because of increased speed limits and higher-speed cruising and cornering, lateral loads have increased dramatically. Automobile suspension systems of the 1990s thus need to satisfy completely different driving patterns than in earlier models, but in the testing department, engineers still apply arbitrary loads to newly developed suspension systems during accelerated durability testing, thereby neglecting significant changes in customer driving-patterns.

Another concern raised by the author was that engineers in the testing department generally had only a limited number of testing samples (or proto-types) on which to conduct their validation tests. It is problematic at best to use test results obtained from a limited number of samples to estimate the population distribution of all mass-produced products operated under real-istic usage conditions. In other words, it is almost impossible to estimate the true population distribution of the products associated with the testing samples at the initial stage of product development. In fact, testing engineers do not even know where on the distribution curve the results of tested samples are located. Unfortunately, traditional testing procedures are usually based on the assumption that if the tested samples passed the testing standards, all products associated with the tested samples would pass the same standards. Similarly, if one tested sample fails in a testing procedure, all products associated with the tested samples are assumed to be failures in the same testing procedure. The author questioned what kind of quality the testing department could assure under such testing standards and procedures.

After two years in the chassis-testing department, the author was transferred to the total-vehicle-testing department at the Murayama Proving Ground, where he was given the assignment of revamping the department and rationalizing testing procedures. "You can use whatever methods you wish, even tough ones, to restructure this testing department," he was told. After his transfer, the author felt that the difference between the proving ground and the technical center (where the design- and chassis-testing departments were located) was as wide as between sky and earth. The author felt he had been demoted to the status of hourly worker. This statement, while not very polite to those already at the proving ground, nevertheless sums up the initial feelings of the author.

The major tasks at the Murayama Proving Ground were to conduct durability tests to evaluate the durability and reliability of prototype automobiles by following long-established testing patterns. Testing engineers looked for signs of problems or failures and worked to correct any that arose. If no problems or failures were detected, testing engineers would assume that the tested prototypes and all associated mass-produced automobiles passed the testing standards. Unfortunately, the results of such testing and validation procedures are very unreliable.

Occasionally, the author asked testing engineers at the proving ground the following question: If the tested prototypes do not have any problems, does it really mean that all cars associated with the prototypes but manufactured by ordinary mass-production processes won't have any problems? In general, the testing engineers did not know how to answer. They assumed that if the prototypes had no problems or failures, the associated, mass-produced cars would have no problems under marketing or customer-usage conditions. Unfortunately, the testing procedures had not been updated to reflect changes in such conditions and were therefore not able to give realistic simulations. Another question arose concerning the testing procedures: if testing engineers usually assigned one of their prototype cars to each testing pattern, how could they know whether differing results were caused by car-to-car variations or by differences among testing patterns? Other questions were raised concerning how long the durability tests should last to ensure that the tested prototypes really passed the testing standards.

Although such questions were raised, they were seldom seriously considered at the proving ground. In fact, the major focus of testing at the proving ground was on how to meet testing schedules in new-car-development processes. No other items were felt to be as important. However, in reality, meeting these testing schedules was not so critical. As it turned out, the major task at this proving ground was neither problem diagnosing/prevention nor meeting deadlines, but simply accomplishing the assigned tests.

## 1.2    EFFORTS IN RELIABILITY IMPROVEMENT

During his time at the Murayama Proving Ground, the author pondered the following question: How can test engineers use only a very limited number of testing samples (i.e., handmade prototypes) to estimate the probability distribution of an entire population (i.e., all mass-produced cars associated with the prototypes)? After doing some research on reliability theories, the author consulted a reliability expert at the University of Tokyo, with whom he developed new reliability theories and testing procedures to update those in use at the proving ground. However, many difficulties were encountered in attempting to apply these new theories and procedures in industry.

The original thinking behind these reliability theories had been to collect as much experimental data as possible in order to estimate the population distribution, and then apply statistical-analysis methods to estimate the risks or failure rates of the population. Theoretically, this approach is unquestionably correct. If time and cost were not major concerns, testing engineers could collect enough data to validate the risks and failure rates of the population with a very high degree of confidence. However, problems arise in the realistic applications of these reliability theories and testing procedures because engineers in testing departments seldom have the resources—in both money and time—to be able to conduct enough tests to collect the required data. Further, new technologies are developed and updated very quickly, and as a result, the reliability data of previous technologies may not be applicable to new technologies. For example, electronic technologies have been widely applied in automobiles and have been updated very quickly. It is already known that the reliability data about old electronic devices cannot be smoothly applied to estimate the reliability of new ones because of different electronic structures, subsystems, and design concepts. Although developing good reliability theories and testing procedures is rewarding, doing so usually takes too much time and money to be a realistic and efficient way to develop future products. As a result, these reliability theories and testing procedures are usually not practical in actual product development. This proved to be the case for the author's theories and tests. Since the reliability theories were not practical, the corresponding testing procedures did not do much good vis-a-vis testing at the proving ground.

At the time, due to competition among major automobile manufacturing companies, the proving ground was under pressure from top management and marketing departments to improve quality and thereby customer satisfaction, with the latter being tracked by such publications as the *Maritz National Car Tracking Study*, Maritz Market Research, USA, and the *J.D. Power and Associates Initial Quality Survey*. These public customer-satisfaction reports became the major focus of car buyers and automobile manufacturers in the global market. Generally speaking, the reports classified customer-satisfaction indexes into the following categories:

- Variation in initial quality of the same model
- Initial quality of the same model in different years
- Initial quality among different models
- Performance deterioration of the same model under realistic usage conditions
- Performance deterioration of the same model in different years, and
- Performance deterioration among different models

After top management mandated quality improvement for all the company's models, quality-assurance departments initiated numerous quality-improvement activities. However, initial-quality and performance deterioration in cars are, in fact, two very different issues. Initial quality problems are related essentially to variations in manufacturing processes, whereas performance-deterioration problems are mostly related to failure or deterioration modes of products or materials. Testing engineers were supposed to determine failure or deterioration of products or materials using test results, so as to solve fundamental deterioration problems. Technology and product development engineers worked very hard to improve the reliability and durability of new products and, in theory, usually got significant results. However, at the testing and validation stages, engineers seldom had the resources to collect enough experimental data to validate theoretically expected results under realistic operating conditions. Thus, it remained unknown whether the efforts spent at the technology- and product-development stages had paid off in reality.

Using traditional quality-engineering approaches, engineers usually first need to determine the root causes for the failure or deterioration of their products or materials, and then solve these failure or deterioration problems. However, determining the root causes is completely different from changing product designs to rectify them. In other words, problem diagnosis is very different from problem prevention. Engineers may be able to determine that the deterioration of a given product is due to some specific environmental factors; however, unless they are able to change the product design to prevent this deterioration problem, they cannot really improve the durability or reliability of the product.

In the author's view, quality improvement is a very different issue from the measurement of product quality or performance characteristics. Quality improvement is achieved by changing product (or process) designs to improve basic functionality. To do this, engineers need to reduce the functional variation of their products (or processes) in the advanced development and design stages so as to improve product (or process) quality in fundamental ways. This is an important, new quality concept. Engineers need to understand that environmental factors together with other noise factors (i.e., disturbance factors) will cause basic product (or process) functions to deviate from their intended ideal conditions. In many case studies of deterioration problems, it is easier to determine the failure mechanisms of products or materials than to fundamentally solve these deterioration problems.

## 1.3   INITIATION OF ROBUST ENGINEERING

Because the basic product (or process) functions are usually negatively impacted by such noise factors as environmental conditions or customer-usage conditions, the author began to search for the most efficient quality-engineering methods to reduce these functional variation problems. Initially, he studied all the methodologies and resources available to determine appropriate quality-engineering methods. However, these approaches were usually so-called "trial-and-error" methods, and required testing engineers to adjust their product- or process-design parameters individually and repeatedly to develop better designs. The author also began studying textbooks and technical papers on robust engineering written by Dr. Genichi Taguchi. However, at that time, the author had some difficulties in understanding the fundamentals of these robust engineering methods (i.e., Dr. Genichi Taguchi's quality-engineering methods), perhaps due to the constraints of a traditional engineering education, which focused on mathematical thinking and proving how the right-hand side of an equation is equal to its left-hand side. Such an education does not encourage students to consider the fundamental meaning behind engineering equations, and most engineering students just follow whatever their professors say without much independent thought, rarely wondering about the basic concepts and philosophies behind the equations in engineering textbooks.

Fortunately, at that critical time, the author got the support of some engineers who had perceived the same engineering problem, and as a result, the Nissan Motor Corporation Reliability Engineering Center was established. One crucial goal of this center was to shift traditional-engineering (firefighting) paradigms to robust engineering paradigms. To this end, the center became awash in robust engineering paradigms, and its engineers tried to popularize such methods throughout the company. Of course, some obstacles were encountered in the popularization of these paradigms, as described below.

One critical problem in the application of robust engineering was that engineers (especially design engineers) seldom had the opportunity to see the actual products or components. They usually applied theories from their engineering textbooks to develop and design new products or to solve engineering problems. For instance, they might use vibration equations to design a suspension system or set the proportional constant for springs in suspension-system calibration in order to meet customer requirements, etc. A chance to see the actual components or products and how they function would have made it much easier for engineers to determine good design-parameter settings and to ensure component quality by specifying technical settings for suppliers.

These were, in fact, very common problems at the technical center. The author thought that these obstacles might explain why the engineers were unable to enhance their technological capability. If engineers are not able to change from traditional-engineering paradigms to robust engineering paradigms, they will not be able to develop high-quality products productively, no matter how much time and resources they spend on testing procedures.

## 1.4 THE REASONS FOR USING ROBUST ENGINEERING METHODS

In the engineering world, many believe that if they can understand all the theories behind one engineering system, they will have sufficient knowledge and capability to develop or design the system. In fact, this fallacy is a critical problem that the engineering society needs to solve as soon as possible. The idea that one prevents serious problems or failures by applying all necessary theories when designing systems is scientific, not engineering, thinking. It stems from the scientist's imperative need to determine the most appropriate principle to explain each natural phenomenon. However, the transition from scientific theory to engineering applications is complex, and so it is that even when applying knowledge obtained from scientific research, engineers are not able to develop very high quality products at reasonable cost. The smooth application of scientific knowledge to the world of engineering requires that adjustments be made. Take the design-review activities for a new product, for example. In general, design engineers think that if the design logic behind a new product is rational and the designed product passes the validation tests, this product will not have any serious problems under actual-usage conditions. However, the author has seen many products that had a very rational design logic fail when used. In general, these failures were due to design-stage engineers underestimating the effects of some critical noise factors that occur under usage conditions. As a consequence, the validation tests that were developed did not adequately simulate realistic usage conditions, and the fact that prototypes passed these tests became meaningless.

After new products are developed and sent to the market or the customer, some problems that cannot be explained by scientific or engineering theories often occur. New-product designs may pass the design-review standards and the design logic may look rational; unfortunately, engineers still do not have the robust technologies to predict or ensure the reliability of new products. Consequently, they are not very confident about the functional robustness of the products under realistic usage conditions.

There were some discussions between Mr. Kaneichiro Imai, the vice-chairman of the Japanese Industrial Education Association, and the author about industrial education in Japan. Many engineers understand the theories of their textbooks, but nonetheless are unable to develop sound engineering systems. Because these engineers cannot develop robust engineering systems, they are not of much value to their companies. One major disadvantage of our engineering education is that we do not teach our engineers how to apply physical laws to develop realistic engineering applications. For instance, as we know, apples were falling before Newton discovered the law of gravity.

To explain the mechanical phenomena in nature, Newton developed such important laws as Reaction Force = Mass x Acceleration. However, there is a big difference between scientific laws and engineering applications. Newton's laws explain the dynamic of a mass, but do not really serve to guide an engineer in developing a good mechanical system. This is the gap that a change in educational emphasis could help alleviate, by having professors teach, for example, how to translate a mass with an acceleration into a useful force that meets the engineering goal. If we cannot solve this fundamental educational problem, we will not be able to productively and efficiently solve our engineering problems.

## 1.5    SUMMARY

Originally, reliability-engineering methods were developed as evaluation/ measurement methods. However, to restructure their technological bases, engineers in manufacturing industries need to shift their paradigms from the traditional evaluation/measurement methodology to robust engineering, thus improving manufacturing productivity. Doing so will be critical to the competitiveness of all manufacturing industries in the fiercely competitive global market. The most critical component in restructuring technological bases is the need to educate engineers and technicians in robust engineering thinking and methods, which then can be applied to product development, new-process development, new-technology development, or even new-market development.

Engineering methods in different fields may be dissimilar, but the robust engineering paradigms in this book can be applied to them all. For instance, the parameter-design (i.e., robust engineering) methods of automobile industries can be applied smoothly to other manufacturing industries, such as electronics, chemicals, etc. The author will illustrate how the concepts of robust engineering can be applied to different engineering applications in the following chapters.

# CHAPTER 2

# The Evolution of Reliability Engineering in the Automobile Industry

## 2.1 INITIATION OF RELIABILITY ENGINEERING

Reliability engineering began during World War II, when electronic military weapons, such as radar, were developed and introduced on the field of battle. In the beginning, the military had very high expectations for these advanced electronic weapons, but their functional reliability was only around 50%, too low to ensure victory in any military engagement. The Defense Department of the United States thought that such low reliability was a severe impediment to the defense of the nation. The first reliability-engineering project focused on the functional reliability of vacuum tubes, which were widely applied in these early electronic weapons. This is the initiation of modern reliability engineering.

Currently, the Defense Department is still the number-one promoter of reliability-engineering techniques. It has published a tremendous amount of material on reliability-engineering techniques and reliability-specification data, such as MIL-STD (**Military Standards** of the United States).

Electric-device manufacturers in Japan started to use reliability techniques to improve product reliability in 1954. One typical example of the reliability problems of electric devices is the unpredictable working life of electric bulbs. Each bulb has a resistance that can convert electric energy into light energy, but this conversion is very nonlinear. In other words, much energy is lost as heat during conversion, producing many damaging side effects that may shorten the working life of the bulb. Electric devices are much more nonlinear and unpredictable in their functional performance and reliability than mechanical products, and it is for this reason that reliability engineering was initiated by electric-device manufacturers in Japan. Generally speaking, the working lives of most electric devices are essentially affected by their initial-failure modes, and so reliability techniques involving screening and detecting were applied to circumvent any potential initial-failure modes that might occur under market or usage conditions.

From these reliability requirements gradually developed designs for reliability methods, reliability-analysis methods, failure modes and effects analysis (FMEA) methods, reliability-testing methods, etc. These methods form the foundation of modern reliability engineering.

Automobile-usage conditions are very similar to those of electric devices. Both need to meet a wide variety of customer requirements and are used under varying operating conditions. It is this wide variation in requirements and usage that has brought about great changes in reliability engineering in the automobile industry since World War II. The history of reliability engineering in the automobile industry is closely associated with changes in the social environment and the development of new technologies. The author is going to discuss the development of reliability engineering in the automobile industry in the remaining sections of this chapter.

## 2.2    MOTORIZATION IN THE 1960s

From the beginning of the 1950s through the first half of the 1960s, major automobile manufacturers were focused on the basic mobility of their products. This was the era of durability and mobility, and it was during this time that Japanese automobile industries began to export cars to the United States. However, because of a lack of knowledge and experience in durability engineering vis-a-vis long-distance highway driving, the durability of the exported Japanese models was much worse than expected.

To accommodate the 1964 Olympic Games, many roads around Tokyo were repaved. By that time, the number of automobile owners had also increased significantly due to growth in the Japanese economy, resulting in increased Japanese mobility (Fig. 2.1). The Meishin (Nagoya to Kobe)

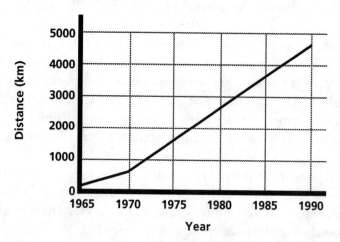

**Figure 2.1** Total length of Japanese domestic highways. (*Japanese Annual Road Survey Report,* The Ministry of Information, Tokyo, Japan, 1992.)

and Tomei (Tokyo to Nagoya) superhighways were also opened at this time, and high-speed performance and safety requirements increased correspondingly. The average performance of Japanese cars increased significantly at this time. Because of the development of durability engineering and the needs of customers, Japanese automobile manufacturers focused their attention on such durability-engineering techniques as maintenance-free design, extended warranty, etc.

As mentioned above, the Japanese industry increased its automobile exports significantly during this time, but found it was very difficult to engineer cars that met foreign weather conditions, road conditions, gasoline specifications, etc., all of which caused many quality problems. Conditions in some of the developing countries of Southeast Asia provide a good example. The roads in most of these countries were rarely paved. Even worse, automobiles in these countries commonly carried loads several times the legal specifications. Tires were often worn, and engines were seldom maintained. It was therefore not surprising that many engine and suspension problems arose, and Japanese automobile manufacturers were forced to develop strategies to overcome these quality problems so as to ensure a market for continued exports to these countries. Another overseas quality problem stemmed from the use of salt to prevent icy roads during the North American winter(Fig. 2.2), a practice that caused rusting of Japanese automobiles exported there. Salt may not only rust through automobile bodies and

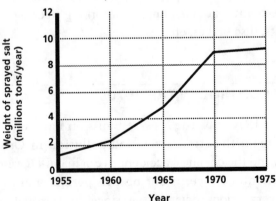

**Figure 2.2** Sprayed salt in North America. (Seiichiro Ide, "Accelerated Tests and Marketing Comparisons for Automobile Body Rusting," Twelfth Reliability and Maintenance Symposium, 1982, held in Tokyo, Japan, and sponsored by the Japanese Union of Scientists and engineers (JUSE).)

damage the body structure, but it may also damage internal components (especially electric and electronic components). Many quality problems were thus faced in the internal electric and electronic systems of Japanese automobiles exported to North America.

After 1965, high-speed performance and durability were no longer critical problems for most automobiles manufacturers, but environmental, pollution, and usage problems (e.g., how to satisfy customer demands) became major issues. Thus, most automobile manufacturers shifted their focus from durability engineering to reliability engineering.

In 1969, laws involving the recall of automobiles for quality or safety problems were issued in Japan and hence became another concern for major automobile manufacturers. The Japanese regulations were based on the automotive-safety regulations of the United States, established in 1966, and specified the conditions under which automobile manufacturers would be forced to recall their automobiles. (This law is still on the books, and many recalls continue to be issued based on it.) At that time, Japan did not have complete regulations for the recall of automobiles, and Japanese automobile manufacturers consequently did not have efficient strategies for handling recall issues.

Generally speaking, reliability and safety issues were the main focus of major automobile manufacturers during this period, and many important safety devices were developed and put into commercial use: retractable fender mirrors, three-point seat belts, impact-absorbing steering columns, dual brake-oil circuits, disc brakes, etc.

## 2.3   EMISSION CONTROL IN THE 1970s

The beginning of the 1970s saw the Muskie Emission Law issued in the United States. In Japan, the strike over lead pollution in Ushigome-Yanagimachi, Tokyo, and the chemical accident in Suginami-ku were examples of the environmental concerns during this period, and pollution problems have been very serious societal issues since. In metropolitan areas in most industrialized countries, air pollution is the most common cause of sore throats and eye irritation for residents. These air-pollution problems are essentially caused by automobile emissions, and as a result, the public has protested strongly, aiming to force automobile manufacturers to reduce the pollution caused by exhaust. The Muskie Emission Law mandated that the emissions of CO, HC, and $NO_x$ be reduced to only 1/10 of the original levels set in 1973. Japan actually executed its own emission control law earlier even than the United States. In 1975, the Japanese Emission Law stipulated that the CO and HC emissions should be reduced to only 1/10 of their original levels, while $NO_x$ emission was officially regulated in 1978(Fig 2.3).

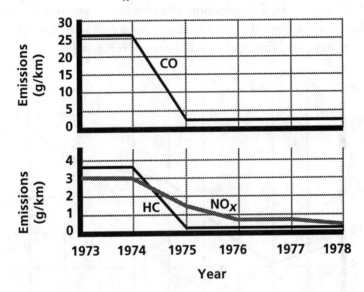

**Figure 2.3** Japanese automobile emission specifications [maximum allowable values. The Environment Agency, Tokyo, Japan, 1975 (CO and HC), 1978 ($NO_x$)].

The primary methods for reducing CO and HC were through lean-burning technologies, thermal reactors, acidification catalytic converters, very accurate control of air/fuel ratio, etc. The basic methods for reducing $NO_x$ were through the applications of exhaust-gas recirculators (EGR). These innovations allowed automobile emissions to be reduced to required levels.

About the same time, no-lead gasolines were introduced to reduce lead pollution, while engine valve seats were produced from various alloy materials rather than cast iron, in order to provide better wear resistance. Through these and other design improvements and new manufacturing technologies, automobile engines were made much more durable than before.

In addition to the pollution problems noted above, the oil crises of 1973 and 1979, along with stricter emissions regulations, greatly increased automobile fuel costs. To compensate, most automobile manufacturers focused their research on fuel economy. One way to improve fuel economy was by improving engine efficiency. It is also possible to indirectly improve fuel economy by reducing chassis weight through the use of high-tensile-strength steel sheets, thus reducing body panel thickness to a minimum while also reducing overall weight. The use of aluminum alloy (to reduce automobile weight) also increased. As a consequence of these advances, the fuel economy of Japanese cars improved drastically, as illustrated in Fig. 2.4.

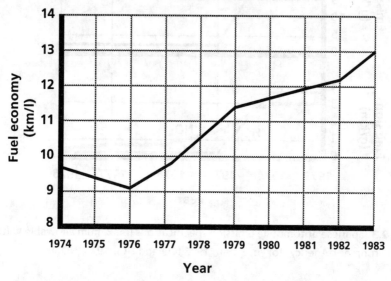

**Figure 2.4** Average fuel economy of Japanese passenger cars; Japanese 10 model type test. (*Environmental White Book,* The Environment Agency, Tokyo, Japan, 1985.)

Another reliability problem was the surface rusting of automobile bodies, which occurred more frequently in Northern Europe and Canada. In 1977, guidelines formally issued in Canada specified that automobiles should have no structure-damaging rust within the first six years of usage, no body-panel-rust-through within the first three years, and no body-surface rust within the first year. Automobile manufacturers in the United States and Northern Europe, where similar weather conditions prevailed, were significantly affected by the issuance of these guidelines, and consequently all major manufacturers focused prodigious energy on solving automobile rusting.

The number of automobile drivers also increased very dramatically during this period, and there was very strong demand for easily operable automobiles. As a result, automatic transmissions and power-steering systems increased in popularity. But high performance continued to be of interest to customers. To meet these demands, many new systems and components needed to be developed within a very short period of time, and more efficient reliability methods, such as FMEA (failure modes and effects analysis), FTA (fault-tree analysis), DR (design review), etc., were correspondingly developed during this period.

## 2.4 ELECTRONIFICATION IN THE 1980s

In the 1980s, to compensate for horsepower loss due to strict emission regulations, turbo-charged and twin-cam engines were developed and commercialized. These engines initiated the era of high horsepower output (per unit of engine volume) and high engine-revolution. Because of the engines' extremely high output, their engine and transmission components suffered greatly increased stress, and new quality problems, such as high-frequency vibration and high-thermal effects, also arose.

To efficiently meet the higher engine output requirements, many automobile systems were electronified; examples include engine fuel-management systems, electronic suspension systems, and electronic automatic-transmission systems. Automobile electronification has indeed been the main focus of major automobile manufacturers since the 1980s. However, as automobiles are typically used under a wide variety of environmental conditions (extremely hot or cold weather, high chassis vibration due to road conditions, rain or snow, muddy roads, electric-wave interference, etc.), their usage conditions are actually much harsher than corresponding usage conditions for ordinary home electronic devices (such as TVs, stereo systems, etc.). Electronic automobile systems were commonly controlled by semiconductor chips, which were not of predictable reliability. Thus, the reliability of electronic automobile systems became the major research topic in the automobile industry in the 1980s. Advances in new manufacturing technologies, new production technologies, new service technologies, etc., have significantly improved the reliability of electronic automobile systems. In particular, the advancement of new welding technologies has prevented many potential reliability problems in electronic systems.

Improvement in engine performance has, unfortunately, increased the number of traffic accidents; so to ensure driver and passenger safety in traffic accidents, advanced innovations—such as air bags, passive seat belts, antilock brakes, four-wheel steering systems, traction control systems—were developed.

Air bags became popular in 1987. The basic reliability requirement of an air bag is that it should inflate automatically and instantly after a head-on collision, but not under normal driving conditions. An air bag system is thus not allowed to have any functional error. Unfortunately, it is very difficult to validate the functional reliability of such a system because of the difficulty in simulating all possible accident or driving conditions. Some automobile engineers have indicated that air bag system reliability requirements are as difficult to meet as those of manned space rockets.

Customer satisfaction was another issue. To further improve customer satisfaction, customer opinions were of major concern for automobile industry administrators. Two new customer trends are (1) more and more women and seniors are driving; and (2) recreational and four-wheel drive vehicles are more popular than ever(Fig. 2.5). Thus it is anticipated that future automobile-usage conditions will vary even more than they have in the past.

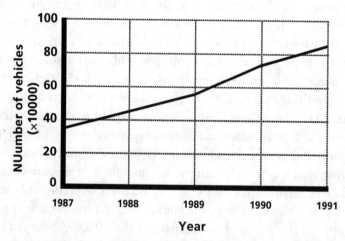

**Figure 2.5** Number of registered recreational vehicles. (*Review of Automobile Industries,* Japan Automobile Manufacturer's Association, Inc., Tokyo, Japan, 1993.)

## 2.5 DEVELOPMENT OF NEW TECHNOLOGIES IN THE 1990s

Because of the growing trade imbalance between the United States and Japan, after 1990, Japanese automobile industries were required to increase their purchases of automobile components from North America. The components purchased from North American suppliers were not always of the same reliability or quality levels as those from Japanese domestic suppliers. To ensure that the production processes at Japanese automobile transplants in the United States were as smooth as those in Japan, new reliability-engineering methods were needed. These will, no doubt, be very different from traditional reliability-engineering methods.

Energy resources, environmental pollution, emissions regulations, fuel costs, hydrocarbon emission specifications, automobile air-conditioner refrigerant specifications, etc., are all concerns for the automobile industry. As mentioned above, emissions regulations in most countries are becoming much stricter, and emissions regulations in Japan are also strict on diesel engine exhaust: $NO_x$, and carbon particles. As all these must be reduced to a minimum, catalytic materials for these pollution emissions are under speedy development.

In California, where emissions regulations are the strictest, the fleets of all major automobile manufacturers will have to include a certain percentage of zero-emission vehicles (ZEV) by 2003. These regulations require that electric cars be developed immediately and put on the market as soon as possible. The automobile industry is also speeding up the development of new power plants—such as solar batteries and motors, methane engines, alcohol engines, etc.—to replace conventional gasoline engines.

The latest technical trends in the automobile industry will focus on the development of new technologies, especially electronic technologies, and will be developed at a very fast pace. These will, no doubt, include navigation systems, antisleep devices, autopilot systems, etc. Most will be commercialized and put into practical applications sooner or later. Advanced traffic-control systems and new materials are also under research and development. To put these new technologies into realistic applications and various market conditions, new reliability-engineering methods that can ensure extremely high reliability requirements must be vigorously pursued. In other words, new quality-engineering techniques need to be structured into the automobile industry (or other manufacturing industries) to serve as the basis of the new technological era.

# Changing Conditions in Society and Strategies for Solving Related Quality Problems

# 3

# Changing Conditions in Society and Strategies for Solving Related Quality Problems

## 3.1    THE IMPACT FROM GROWTH IN NEW TECHNOLOGIES

Reliability engineering in automobiles consists of two major areas, as illustrated in Fig. 3.1: (1) improvement of initial quality level, and (2) maintenance of said quality level for the product's entire operational life. Currently, quality problems are a major concern throughout the automotive industry, and conventional reliability-engineering techniques are widely applied to ensure the functional reliability of automobiles. Reliability engineering is essential in both ensuring the basic functions of an automobile and maintaining these functions for its entire operational life. Automobiles are driven under widely varying environmental and operational conditions, yet we expect their basic functions (e.g., mobility) to vary as little as possible. We also expect the deterioration of these basic functions to be minimal throughout the operational life of the product. So, stable function and minimum deterioration are required to develop highly reliable automobiles.

The author defines reliability as follows: the total technological capability to develop products that can both perform predetermined basic functions and withstand associated deterioration of said functions throughout the product's operational life. This definition of reliability highlights the major challenges (i.e., improving and maintaining basic functions) currently seen in quality engineering. Unfortunately, the basic concepts of traditional reliability engineering are somewhat misunderstood and tend to focus on measurement techniques for product reliability, rather than on how to improve actual reliability. In other words, traditional reliability engineering focuses not on how to improve the quality (and thus reliability) of a product (through quality-by-design techniques), but on how to measure defects and analyze failures in products. As a result, traditional reliability engineering is inefficient in improving the quality levels of newly developed products.

In the past, reliability engineering was used to improve traditional technologies, which were primarily based on mechanical systems. This was possible because mechanical technologies usually grew slowly; the accumulated experience and knowledge of the mechanical technologies of a preceding generation could be efficiently applied to update and improve the basic functions of a new generation of technologies. Generally, when different generations of technologies overlap, the tremendous amount of testing data available for old technologies can be reanalyzed and reapplied to improve

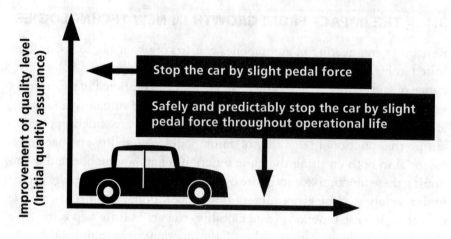

**Figure 3.1** Initial quality assurance and reliability assurance of automotive brake system. Reliability is the improvement of initial quality level plus maintenance of this quality level throughout the product's operational life.

the basic functions of the new technologies. It is always possible to find a better way to improve the functional stability of current mechanical technologies. Thus, traditional reliability-engineering techniques, though inefficient by current standards, played an important role in the improvement of traditional mechanical technologies.

In the era of mass production, products were manufactured in very high volume, but with few varieties. In these mass-production systems, product designs were seldom changed unless absolutely necessary, and quality-engineering activities were primarily focused on improving existing production processes rather than on product designs (because product designs were seldom changed unless the models changed). In addition, the quality requirements of customers were not as demanding, because living standards in years gone by were lower than at present. Therefore, at that time, accumulated technological data, experience, and lesson manuals could be applied to develop the next generation of technologies.

As current living standards are much higher than, say, 20 years ago, people tend to require more to fulfill their personal needs and wants. As a result, the demands placed on owning an automobile in Japan have increased dramatically, as illustrated in Fig. 3.2. For the same reason, a much wider variety of quality and performance requirements are demanded by a much larger group of customers. Thus, the number of automobile models and their functions have grown. To meet these increasingly varied requirements, current production processes have been updated to be of the high-variety—low-volume type. Product liability and customer satisfaction have become the two most critical factors for the growth, indeed the survival of manufacturers. To grow (or survive) in this highly competitive global market, manufacturers need to rethink how to engineer quality into their products and also how to restructure their organization to achieve this objective efficiently.

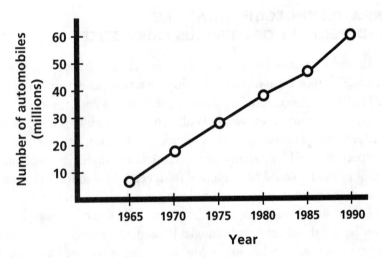

**Figure 3.2** The total number of automobiles in Japan. (Japan Automobile Manufacturers' Association, Inc., Tokyo, Japan, 1992.)

To meet the challenge posed by both increasingly varied customer requirements in product quality and performance and by ever-changing conditions in society, manufacturing companies need to improve their technological capability, especially in the fast-growing areas of electronic and IC (integrated-circuits) technologies. Currently, these are being developed at an astounding rate. In view of this, the author would like to introduce a new term, technological half-life, to illustrate how fast new technologies are being developed.

Assume that the scale of a given technological leading index ranges from 0 to 10, with 0 denoting outdated or noncompetitive technology, 5 denoting current technology, and 10 denoting advanced and leading technology. The time required for a leading technology to drop from 10 to 5 in its technological leading index is defined as its technological half-life. The technological half-lives of typical mechanical technologies are around 7 to 8 years. By comparison, the technological half-lives of typical electronic technologies are only about 2 years, and those of IC technologies are only around 1 year. These figures clearly show that new technologies based on electronic and IC systems are being developed at an alarming rate.

## 3.2 NEW TECHNOLOGIES THAT ARE INDEPENDENT OF PREVIOUS EXPERIENCE

The evolution of automobile technologies serves well to illustrate why previous experience cannot be applied to developing a new generation of technologies. The average model life of an automobile (from concept initiation to end of sales) is about seven years. Traditional automobile technologies were essentially based on mechanical-engineering systems; thus, the accumulated experience and knowledge of mechanical technologies associated with a previous model could be reapplied to develop new models (because of the slow pace in mechanical-technology growth). Existing technologies could be transferred to the new model up until the end of current model life without losing significant technological competitiveness.

However, automobile technologies are now being developed at a much faster rate. Take fuel injectors of automobile engines, for example. Carburetors, originally completely mechanical, were used in automobile engines to vaporize fuel and mix it with air at a predetermined ratio. However, because mechanical carburetors were not sophisticated enough to make the very sensitive adjustments in air/fuel ratio required to meet the increasingly strict requirements of engine performance and emission regulations, electronic fuel injectors had to be applied instead. Similarly, many electronic control systems were—and continue to be—added to automobiles at a very fast rate.

If automobile-development engineers neglect trends in new automobile technologies, they may select systems that will be outdated before the newly developed models are actually delivered to market, and the applied new systems will be non-value-added items for the newly developed automobile models. In other words, no technological advantage is actually gained if a new technology is not delivered to markets smoothly and quickly.

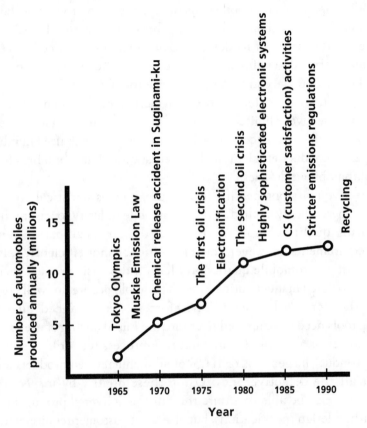

**Figure 3.3** Number of automobiles manufactured annually in Japan. (Japan Automobile Manufacturers' Association, Inc., Tokyo, Japan, 1992.)

New technologies are developed to meet the increasing and ever-changing requirements of society, many in new fields in which there is no previous experience or experimental data. As a result, previous experience, knowledge, or lesson manuals are becoming meaningless for the development of new technologies. The associated engineering tasks for the development of new products are becoming much more complicated because of an increasing number of unknowns. Currently, many new technologies are, in fact, based on some knowledge that engineers have never encountered. How to ensure the functional stability (i.e., robustness) of the newly developed technologies in the very early stages of product development has become a new technological challenge for engineers.

Since 1990, Japanese automobile companies have been producing about 13 million automobiles annually (Fig. 3.3). This number naturally includes an extremely wide variety of models, some of which are produced in very large volume. Rarely do engineers have access to sufficient prototypes to simulate all possible usage conditions of these automobiles at the very early stages of the product-development process. In most cases, development engineers can only use a very limited number of prototypes to conduct validation tests for customer-usage conditions. How to use a very limited number of prototypes to conduct efficient quality assurance activities is a big challenge for automobile industries.

Downstream quality-engineering techniques, such as statistical process control, have been widely applied in the automobile industry to ensure that manufacturing quality characteristics meet design specifications and customer requirements. However, these activities cannot efficiently solve the fundamental automobile quality problems if the basic product technologies or associated manufacturing-process technologies are not well engineered at the very early stages. Therefore, more advanced quality engineering tools need to be applied to ensure the initial quality of the product technologies or associated manufacturing-process technologies. It is possible to conduct inspections of 100% of all manufactured products and screen out all the defective ones. Using these quality-by-inspection approaches, we may be able to ensure that all the delivered products not only fall within design specifications, but also meet customer requirements. However, this involves scrapping all defective products and manufacturing additional products to replace them. These enhanced inspection activities also require more quality-engineering resources, for training operators and purchasing inspection equipment. All in all, such increased inspection activities result in increased costs, making the products less competitive.

To solve downstream quality problems efficiently and fundamentally, we need to focus our attention on solving the quality problems of product technologies or associated process technologies at a very early product-development stage, that is, at the technology-development stage. Because of fierce competition, most companies want to develop and deliver their new products as quickly as possible, so the lead-time-to-market period for a new product is much shorter than before. If we cannot ensure the initial quality (i.e., the most upstream quality characteristics), we will never have enough time to solve all the associated downstream quality problems at later stages. Development engineers will not have the time or resources to ensure new product quality unless quality problems have been solved at the technology-development stage.

## 3.3 TECHNOLOGY-DEVELOPMENT ENGINEERS AND QUALITY PROBLEMS

Technology-development engineers must concentrate on solving fundamental quality problems in their new technologies. However, such efforts won't mean much if engineers use traditional scientific research to study the cause-effect relationships between the functional variation and downstream noise factors of new technologies instead of making the technologies robust against these downstream noise factors. Generally, even though these research activities are not really productive in industrial applications, it is very difficult to convince engineers not to conduct them in their new technology-development projects, because they were all trained to do so in engineering schools. Engineers need to anticipate the vital few downstream noise conditions and conduct robust-engineering activities to desensitize their new technologies against these anticipated downstream noise factors at the development stage. It is too late to conduct quality-engineering activities after quality problems actually occur at downstream stages. In fact, if they do not do research in anticipation of downstream noise factors using robust-engineering methodologies, development engineers cannot know what quality problems may occur or when. It takes time to determine the root causes of downstream quality problems. Even worse, in many cases the causes stem from environmental factors such as ambient temperature and humidity or deterioration factors such as wear and degradation. These noise factors are very difficult and expensive to control through traditional design activities. It takes even longer to discover the appropriate engineering actions that might compensate for these quality root causes than it does to determine what they actually are. During this period of problem solving, many customers become dissatisfied and even angry with the products, and eventually the manufacturing company risks being the bigger loser because of decreasing market share and increasing customer dissatisfaction.

If we don't solve quality problems early in the product-development process, many associated downstream quality problems may occur later. Fire-fighting downstream is a very inefficient way of attempting to solve quality problems. Using downstream quality-engineering activities, the efficiency of the product-development process will be low; consequently, development engineers (in Japan) find they must work overtime to compensate for the inefficiency of their product-development processes. Too much overtime may negatively affect the efficiency of these engineers and further impact the downward spiral in quality. These causes and effects may form a never-ending negative cycle. In fact, engaging in more downstream quality-engineering activities may even increase quality problems in later downstream stages.

The root causes of most quality problems are related to the functional variation of products or processes. It may not be easy for engineers to shift their old quality paradigms (downstream "fire-fighting" quality engineering) to new quality paradigms (upstream robust-engineering activities) immediately. However, the most efficient and fundamental point at which to solve quality problems is at the most upstream stage of the product-development process, that is, at the technology-development stage.

# CHAPTER 4

# Quality Problems
# and
# Action Strategies

## 4.1    THE ENDLESS STRUGGLE AGAINST NOISE FACTORS

As mentioned in previous chapters, most quality problems are related to variation caused by noise factors. Many engineers believe that if they understand all the theories behind a product and apply these theories to its design, no serious quality problem should arise. Unfortunately, the manufacturing processes of any designed product undergo variation caused by such noise factors as operator error, material nonhomogeneity, etc. Because of these noise factors, the products may have some functional variation, and the associated downstream quality characteristics of the products may deteriorate. Design and process engineers must routinely do battle with noise factors in order to improve the quality of future and existing products.

It is more cost and time efficient to desensitize products to noise factors than to eliminate them. For example, ambient temperature always varies with seasonal changes, which may cause viscosity variations in automobile hydraulic system oils (e.g., braking oil, power-steering oil) and thereby affect their performance. It is almost impossible to keep ambient temperature constant. Similarly, we can conclude that it is impossible or too expensive to eliminate even a small number of noise factors from ordinary industrial products. Even if we were able to afford the cost of conducting the necessary research to identify all noise factors, we still would not be able to eliminate them all. Although engineers work very hard (even overtime, frequently), they seldom can solve the fundamental quality problems of their products.

On the other hand, new technologies are developed at a fast rate. The productivity of all product-development processes of manufacturing companies must be improved to keep up with the rapid growth in new technologies. Otherwise, we won't be able to fundamentally and efficiently solve the quality problems of our products.

## 4.2    THE FAST ADVANCEMENT OF NEW TECHNOLOGIES

Customer requirements are becoming increasingly diversified and sophisticated. To satisfy consumers, new technologies (essentially based on electronic-engineering systems) are being rapidly developed by manufacturing companies. For example, to meet the high-performance requirements of customers, numerous electronic-control systems are being developed and added to automobiles. With the development of these new systems, the responsibilities of automobile engineers are heavier and their tasks more intricate than before. In fact, engineering tasks are increasing at a rate faster than anyone could have predicted. Engineers can seldom refer to their previous experience in the development of new technologies because new technologies are usually developed from new engineering fields in which little experience or testing data have been accumulated. Many engineers work hard to conduct the increasing number of engineering tasks caused by the fast advancement of new technologies; however, few engineers can keep up.

The transfer of production from Japanese plants to overseas transplants causes another quality problem. Most Japanese overseas transplants adopt the basic product designs from the engineering headquarters in Japan, then purchase components and materials locally. Unfortunately, these overseas plants seldom can purchase components or materials that are standardized or categorized to the same detail as those made in Japan. In general, the quality of components or materials purchased overseas is not the same as those made in Japan.

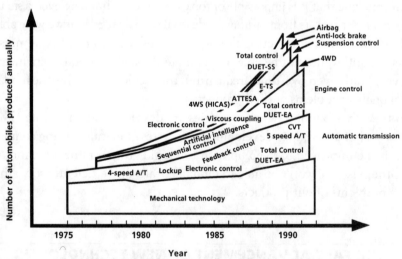

**Figure 4.1** The development of automobile technologies.

| | |
|---|---|
| **SS**=suspension/power-steering | **4WD**=4-wheel drive |
| **EA**=engine/automatic transmission | **CVT**=continuous variable transmission |
| **DUET**=development of united engine/transmission | **ATTESA**=advanced total traction engineering system for all |
| **ECCS**=electronic concentrated engine control system | **E-TS**=electronic-transmission system |

As mentioned above, at the new-technology-development stage, there is little previous experience or testing data to which one can refer. Thus, new quality-assurance methodologies need to be developed to ensure the functional stability of newly developed technologies even at the early stages of product development. Ensuring the quality of newly developed technologies will be a new challenge for those in the robust-engineering field.

This challenge needs to be met so that quality engineering can catch up with the growth in new technologies. Because of this fast advancement of new technologies, the number of testing items and testing specimens are increasing significantly. Without a corresponding increase in testing equipment and resources, most tests on new technologies have to be conducted under a very tight schedule. In other words, if we keep applying traditional quality-assurance methods, the possibility that key quality characteristics will be overlooked will increase. Eventually, more and more customers will be dissatisfied, and the manufacturing companies may be the bigger loser because of decreasing market shares. If we do not change current quality-engineering activities, the efficiency of total-product-development processes will decrease vis-a-vis the fast advancement of new technologies.

## 4.3  SCIENCE AND ENGINEERING

Most quality problems are caused by the variation effects induced by noise factors. Surprisingly, most engineers do not completely understand variation. Scientific research activities, in which variation is not a major concern, are very different from engineering activities. Unfortunately, many engineers confuse engineering activities with scientific research.

In the world of science, the primary objective is to determine the most appropriate principle to explain the phenomena seen in nature. Discovery, new functions, and knowledge are the major concerns of scientific research. On the other hand, in the engineering world, all products experience functional variations from the ideal, regardless of how well they are manufactured. How to make the actual functions of products approach their ideal functions under various noise conditions should be the major concern of engineering research activities. In fact, in the engineering world, there is no so-called "correct method." The engineering method that can minimize the functional variation of products cost efficiently will be the most appropriate method. Profit and practicality are the most important issues in the engineering world.

It is therefore clear that the objectives of engineering are very different from those of scientific research. Unfortunately, many engineers, even those long involved in product manufacturing, still do not quite understand the problem of variation. In the author's opinion, the primary reason for this is the fact that most have been educated in engineering schools where variation is seldom discussed. In this type of educational environment, most engineering students learn only the fundamentals of science, not the variations inherent in the real world. Without further in-house training in dealing with variation, these engineering students will be ill-equipped to solve those variation problems that arise in the real world.

Engineers ignorant of variation have difficulty conducting engineering activities that reflect actual conditions. They often misunderstand the fundamental functions of engineering systems and tend to use the average values of engineering characteristics to analyze their engineering systems, with little thought given to variation effects. As a result, their products may not be robust enough to survive in a world full of variation.

The author heard the following statement while teaching a robust-design course:

> I am not going to abandon a design just because of a large variation in its quality characteristics. A large variation does not affect the quality of a product significantly. I am therefore going to apply the mean value of the testing data to evaluate the performance and quality of the design, without considering variation effect.

Examination of this statement will show how impractical it is. The contents, dimensions, and manufacturing conditions of all products experience some degree of variation caused by myriad noise factors. Assume that the target value of the objective quality characteristic of a given product is 100 units and that we have two suppliers for this product. These two suppliers each provide three samples, as shown below.

| Supplier A: | 98 | 100 | 102 |
| Supplier B: | 70 | 100 | 130 |

Which supplier's products should an engineer use? What range of variation is considered acceptable? These are questions many engineers are unable to answer. In this simple example, if we consider only the mean values of the test data, we would assume that supplier B's products are as good as A's. However, notice that this isn't really so, since supplier B's products have far greater variation than supplier A's. The point is, we should not solve engineering problems by only considering the mean values of the objective quality (or performance) characteristics without the associated variation effects. Otherwise, the engineering problems may not be fundamentally solved.

In total-product-development processes, any design and manufacturing activity may cause variation effects on the objective functions of products. If engineers neglect the causes of these variation effects, they will never be able to fundamentally solve engineering problems. Unfortunately, it is still difficult to persuade engineers to accept the concept of variation and consider variation effects in their engineering activities.

## 4.4   VARIATION

Mr. Tsuchiya, the executive managing director of Fuji-Xerox, made the following statement:

> One major roadblock in the development of new technologies at most manufacturing companies is that the downstream variation effects of newly developed technologies are not taken into consideration at the technology-development stage. These activities are far remote from the real world. Generally speaking, engineers at technology-development departments tend to apply deterministic (i.e., mean-value-only) methods to solve their engineering problems, without taking downstream variation effects into consideration.

Another relevant quote was made at a presentation by Mr. Kunio Isobe, a consultant from the Japanese Administration and Management Research Institute:

> Those activities that aim to discover principles which describe natural phenomena are called scientific research. Applying these principles, discovered by scientific-research activities, to the manufacture of commercial products is called engineering R&D. Unfortunately, most

*engineering R&D activities do not take the variation effects of the real world into consideration. Generally speaking, engineers spend 20 years in an educational environment where variation is seldom discussed and therefore do not understand the variations of the real world. It is not, then, surprising that most engineers do not take variation into serious consideration in their problem-solving processes and that their engineering systems are sometimes not robust enough to withstand these variations. In the real world, variation effects are unavoidable; engineers need to take variation seriously. Since most engineers are educated in an educational environment that ignores variation, they tend to think about their engineering problems using average (i.e., mean) values of objective quality (or performance) characteristics. In other words, they usually neglect the associated variation effects. In engineering education, variation is usually neglected; thus, engineers from this type of environment tend to neglect the variation effects of the real world.*

*As most engineers do not understand the variation effects of the real world, it is important to change their attitudes from mean, value-oriented thinking to variation-oriented thinking. Again, in traditional technology-development processes, the variation effects of downstream noise factors are seldom considered.*

As the above quote states, most engineers do not understand the fundamentals of variation processes. They tend to think about their engineering problems only through mean values of objective quality (or performance) characteristics, while neglecting variation effects.

Variations can be categorized into three types: part-to-part variation, external/environmental variation, and deterioration variation. For example, variations in ambient temperature and environmental humidity are two common causes of functional variation in automobile engines. Some examples of deterioration variation in automobiles are wear due to cycling motions between pistons and cylinders, fatigue due to chassis vibration, the fading of paint, etc.

Anyone who has been involved in the manufacturing process knows that it is unwise to neglect the part-to-part variation of products due to imperfect manufacturing. Yet in the experience of the author, variation effects caused by environmental and deterioration noise factors usually have more impact than those caused by part-to-part variation effects and are therefore usually the major root causes of customer dissatisfaction. Some engineers have already come to understand their importance. However, they usually adjust product functions to compensate for variation effects caused by these two types of noise factors (instead of desensitizing their products against them). Compensating through adjustment of basic product functions is not really an efficient way to make products robust. The author calls this type of adjustment method "fire-fighting" quality engineering. Compensating for the variation effects caused by environmental and deteriorating noise factors usually makes the engineering systems much more complicated, and the efficiency of such fire-fighting quality engineering is usually low.

In summary, environmental and deterioration noise factors should be desensitized through robust-design methodologies at the very early stages of the product-development process. It is too late and too expensive to make products robust at downstream manufacturing stages.

## 4.5　EMPTYING AND RESTRUCTURING OF TECHNOLOGY BASES

In this section, the author will examine the evolution of automobile-development processes in order to explain the root causes of the emptying of technological bases.

At the beginning of automotive history, a few automobile enthusiasts devoted themselves to the basic functions (e.g., driveability) of automobiles, then to designing and manufacturing their automobiles themselves to include these basic functions. Finally, they conducted validation tests against

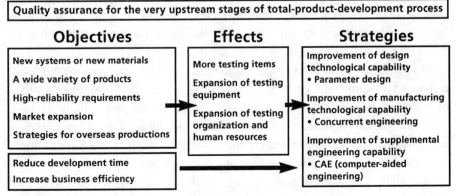

**Figure 4.2** Major issues in a new-product-development process.

the basic functions of their automobiles. In this type of development process, each individual in the automobile industry needed to know many basic automobile functions.

The demands on automobiles first increased gradually, then explosively, and automobile-manufacturing companies grew accordingly. How to improve the productivity has been, therefore, a very popular topic.

Today, one popular way to improve productivity is to improve the efficiency of the total-automobile-development processes (Fig. 4.2). Very sophisticated labor-division methods are being developed and applied to achieve this objective. More people than ever are now involved in the total-development process. Not all of them are automobile enthusiasts. For the above reasons, the range of the working tasks of each individual in the automobile industry has narrowed over the years, and most automobile engineers now conduct routine tasks and work according to the predetermined automobile-development processes of their companies.

Most engineers are therefore only familiar with their routine assignments, very narrow areas of the automobile-development process. Few automobile engineers have a profound knowledge of the total automobile-engineering system. This means they can discern only very superficial phenomena, and are unlikely to think in a systematic way. Also, because they are engaged in routine fire-fighting tasks, they cannot think about how to improve fundamentally the basic functions of their engineering systems. As most cannot probe deeply into the fundamentals of their engineering systems, their technological capabilities are decreasing, which in turn is emptying the technological bases of (Japanese) manufacturing companies.

Although most (Japanese) engineers are working hard, frequently overtime, to compensate for these inefficiencies, unfortunately they still cannot fundamentally solve these quality problems. The primary reason is that they do not have sufficient technological capability.

In the author's experience, management or administrative methods do not stop this emptying of technological bases. The only way to solve this problem efficiently is through the restructuring of said technological bases. This entails an engineering re-education program to show engineers not only the importance of variation reduction, but also ways to desensitize their engineering systems against variation effects caused by downstream noise factors. Without such action, the technological capability of engineers will continue to diminish.

In summary, there are three major objectives in improving the competitiveness of a manufacturing company:

1. To improve the efficiency of the total-product-development process and thus reduce the span from lead time to market of new products.
2. To keep engineers from their fire-fighting routine.
3. To restructure technological bases (an engineering re-education program on variation reduction).

One time-proven way to achieve these three objectives efficiently is through the robust-engineering methodologies developed and promoted by Dr. Genichi Taguchi, especially his parameter-design methodologies.

## CASE STUDY 1: STATIC WELDING STRENGTH AND LASER-WELDING TECHNOLOGY

*Note:* All case studies in this book emphasize how to bring variation into engineering problem formulations. No engineering equations are included in these case studies because they usually cause engineers to focus on the mean values of the objective characteristics and thus to neglect variation effects.

To give readers an initial picture of what is involved in the development of robust technology, a successful application of parameter design to laser-welding technology at the author's company is introduced here. This case study was published in the *Japan Study Mission*, sponsored by American Supplier Institute, Allen Park, Michigan, U.S.A. Parameter-design (i.e., robust engineering) methodologies will be covered more fully later, in Chapter 5.

When we talk about welding quality, we often focus our attention on welding strength, which can be classified into two major components: static welding strength and cycling fatigue strength. If we apply an extremely large static load to a welded part, cracks should occur first on the surface of original material, not the welding surface. In other words, the static welding strength should be greater than the original material strength. In addition to destructive static loads, small cycling loads may cause fatigue in a welded part, especially at the welding location. Fatigue failure modes depend on how cycling loads are applied to the welded part and may differ considerably from each other. However, most welding engineers would agree that if static welding strength is great enough, the possibility of cycling fatigue problems will be very low; this is because cycling fatigue strength is usually correlated to static welding strength to some degree. In addition, static welding strength is usually more important than cycling fatigue strength. Therefore, to simplify this study, the author will focus on static welding strength.

The welding department of the author's company wanted to develop a generic procedure to ensure the functional robustness of their laser-welding technology. The welding engineers were not initially familiar with para-meter-design methodologies. Surprisingly, they were eager to adopt these methodologies to develop robust laser-welding technology.

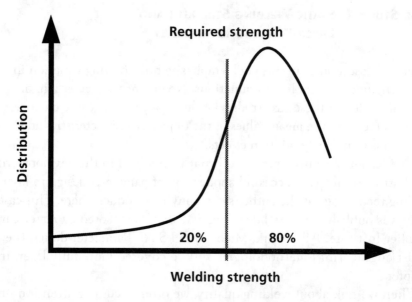

Figure 4.3 Welding strength distribution of nonrobust welding technology.

The distribution of welding strength on testing plates processed by a nonrobust laser-welding machine, as explained to the author by welding engineers, is illustrated in Fig. 4.3. This was the first step in the development of robust laser-welding technology. Traditionally, spot welding is used in plate–plate welding. However, because of higher welding-quality requirements, the welding department wanted to adopt laser-welding technology. One advantage of laser welding is that the welding area is much smaller.

Thus, the welding area can be placed very close to the plate edge to reduce fringe length, as illustrated in Fig. 4.4.

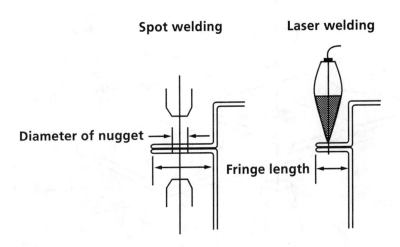

**Figure 4.4** Fringe lengths in spot welding and laser welding.

Compared to spot welding, laser welding needs less welding area and thus provides greater flexibility for design engineers, a primary reason for its adoption. The second reason for using laser welding is that spot welding usually generates numerous welding spots on the plate surface, which affect plate appearance (Fig. 4.4). Further, as plates processed by spot welding require wider fringes, they are usually heavier and have more plate–plate interferences. However, one disadvantage of laser-welding technology is that welding strength is not as strong as in spot welding. Welding engineers had applied 20-mm-long laser welds in welding experiments and found that the failure rate was around 20%, as illustrated in Fig. 4.3. The engineers were not able to improve the figures, despite numerous attempts. Before this robust-engineering project, the welding engineers assumed that the welding strength of a 20-mm laser weld was equivalent to that of a spot weld of a certain diameter (e.g., 8-mm), as illustrated in Fig. 4.5.

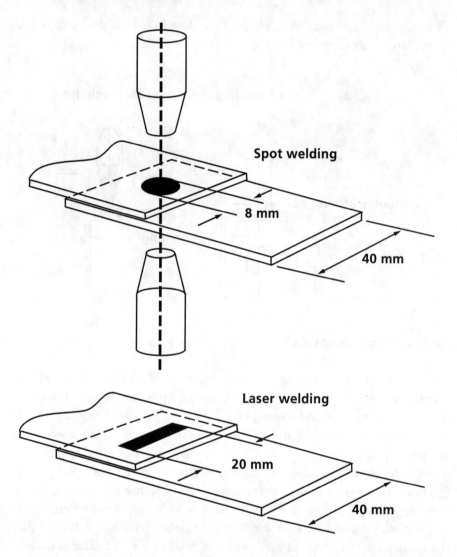

**Figure 4.5** Strengths of spot welding vs. laser welding. The strength of a 20-mm laser weld was found to be equivalent to a 8-mm spot weld.

After reviewing the engineers' development procedure in laser-welding technology, the author concluded that the current technology would fail because of an inappropriate measurement characteristic: equivalent welding strength. The engineers were applying the equivalent welding strength in spot welding to measure the mean value of static welding strength in laser welding, without taking variation effects into account. This measurement characteristic was from traditional engineering thinking. (Most engineers continue to conduct their engineering projects in this fashion.) However, from the viewpoint of robust-engineering, the mean value of static welding strength is only the second most important issue. The mean value of static welding strength can be easily adjusted in downstream stages (e.g., product-design stage) through calibration of the total welding length of the product. Using only the mean value of static welding strength is not a very efficient approach to making welding technology robust against downstream noise factors. This conventional engineering thinking (i.e., mean-value-only thinking) was based on meeting downstream engineering characteristics, not on solving the fundamental engineering problems, which involved variation effects caused by noise factors. Mean-value-only engineering approaches can only be applied in product-development or more downstream stages. In the new-technology-development stage, variation reduction should be the major concern; engineers need to desensitize the basic functions of new technologies to variation effects caused by downstream noise factors.

Japanese engineers worked hard to develop high-quality, competitive products for the global market, and there is no doubt as to the economic power of Japan and the quality of Japanese products. Currently, however, Japanese products are essentially developed through product-oriented types of development processes, in which product quality is achieved through both downstream assessment of individual products and tuning of processes in an attempt to meet customer requirements for quality. Such development processes are inefficient and cannot keep pace with the rapid advance of new technologies. In other words, Japanese manufacturing companies still have no reliable processes for the development of robust technologies. The only way to make new-product-development processes more efficient is to bring robust-engineering upstream, ahead of the new-product-development stage (the new-technology-development stage). The author calls this type of development robust technology development, the meaning of which is somewhat different from traditional technology development. In traditional technology development, the functional robustness of new technologies is seldom considered. Furthermore, the traditional stance is that, due to innumerable and unpredictable downstream noise factors, it is usually too time consuming to fully deploy newly developed technologies into mass production. In contrast, in robust-technology development, the functional robustness of new technologies is the major concern.

Let us return to the welding case study. If the welding lengths of various products are all 20 mm, it is possible to use engineering resources to calibrate various welding conditions and thereby to maximize static welding strength. Unfortunately, the welding lengths of products vary. The objective of robust-welding-technology development is a flexible welding technology capable of processing a wide range of welding lengths. Before developing this robust-welding technology, most welding engineers believed that if they could determine the optimal welding conditions for 20 mm welds, the same welding conditions would apply to other welding lengths. This assumption is based on traditional engineering thinking and is very misleading. To avoid potential pitfalls, the author suggested to the engineers that they apply the proportionality between welding strength and welding length to measure welding functionality, then attempt to reduce variation in this proportionality, thereby reducing the functional variation. Ideally, the proportionality between static welding strength and welding length should be a constant, even under various noise conditions (e.g., different operators, ambient temperature). Experiments were therefore conducted to determine how to make the real function proportionality as close to constant as possible within a certain welding-length range.

After this robust welding technology was developed, the director in charge was so impressed with the power of robust-welding technology that he asked to be shown dynamic-type robust-engineering methodologies in robust-technology development. Indeed, such methodologies are difficult to understand as engineers focus on downstream quality characteristics instead of considering the basic (i.e., fundamental) functions of their engineering systems and products.

Upon the director's request, the author initiated a training program by asking the following question:

> If the static strength of a 20-mm weld is 400 kgf, what will be the
> static strength of a 30-mm weld?

The engineers' immediate response of 600 kgf is an answer based on mean-value-only thinking. In fact, the theoretical answer to this question is indeed 600 kgf. However, if we measure the static welding strength of a 30-mm weld, we find that the real static welding strength varies significantly around

600 kgf. This illustrates that scientific and engineering theories seldom take variation effects into account. Good engineering involves applying engineering methodologies to produce the realistic output of an engineering system that is as close to the ideal output as possible under various noise conditions. The fundamental (i.e., basic) function of a welding machine is to translate electrical energy into thermal energy, which is used to melt the welding material and thereby to connect the metal pieces. This is, in short, a kind of energy-transformation process. If the energy is homogeneously transformed, welding strength will vary only minimally. Or, if every unit of welding length gets the same thermal welding energy, the welding strength per unit of welding length will experience little variation. In this case study, the objective was to develop a welding technology that could provide homogeneous energy transformation, from electric into thermal, on the target material.

One very important issue for the restructuring of technological bases is to change engineers' paradigms from product-oriented thinking to technology-oriented thinking. Product-oriented thinking focuses on improving the quality characteristics of a specific product (e.g., a fixed welding length of 20 mm). However, at the technology-development stage, we do not deal with specific products. Instead, the goal is to develop a generic (e.g., welding) technology for future products not envisioned at the new-technology-development stage. Thus, welding length needs to be treated as a variable, not a constant. Ideally, welding strength should be linearly proportional to welding length. If this generic welding technology is robust, we will be able to apply it to various future products. This constitutes the central theme for robust-welding-technology development.

After viewing Fig. 4.6, most of the welding engineers understood the theme of robust-technology development and immediately conducted experiments under the guidance of the author. The testing plates were 40 mm in length. However, three welding-length settings were used: 10 mm, 20 mm, and 30 mm. Welding length was the input signal factor, and welding strength, the corresponding output response. In addition to this signal factor, the engineers also applied some noise factors (e.g., the gap variation between upper and lower plates, the distance between laser lenses and plates) in the experiments to simulate the variation effects in the laser-welding technology. The relationship among welding length, welding strength, and noise effects is illustrated in Fig. 4.6. The objectives in this welding-tech-

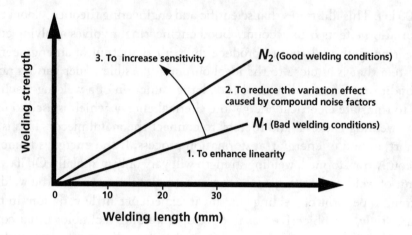

**Figure 4.6** Proportionality between welding strength and welding length, together with associated noise effects. This proportionality is the engineering basis behind robust-welding-technology development.

nology development were (1) to enhance linearity between welding length and welding strength; (2) to minimize the variation effects caused by noise factors; and (3) to maximize the proportionality constant between welding length and welding strength. If all three objectives were achieved, this welding technology would be very efficient in converting electric energy to thermal welding energy. Further, if the proportionality constant were increased, the welding strength per unit of welding length would increase accordingly. As a result, we would be able to reduce the total welding length (or even electric energy per unit of welding length) and thereby reduce manufacturing cost.

In this case study, the engineers combined several noise factors into a two-level compound noise factor, $N$. They also assigned all controllable factors of the welding machine to an $L_{18}$ inner orthogonal array and then conducted the 36 (= 18 × 2) runs of experiments to collect the data shown in Table 4.1. In creating robust-technology, never delete experimental data that looks abnormal, as this neither solves the variation problems that occur in the real world nor improves the basic functions of target engineering systems. After calibrating controllable factors on the laser-welding machine, the engineers went on to make the welding technology so robust that it has already become an important part of the technological base of the company.

TABLE 4.1 Test results for robust-welding-technology development (in kgf)

| Experiment no. | Noise | Welding Length | | |
|---|---|---|---|---|
| | | 10 mm | 20 mm | 30 mm |
| 1 | $N_1$ | 466 | 512 | 814 |
| | $N_2$ | 452 | 538 | 832 |
| 2 | $N_1$ | 220 | 500 | 682 |
| | $N_2$ | 381 | 530 | 784 |
| 3 | $N_1$ | 433 | 534 | 704 |
| | $N_2$ | 220 | 119 | 284 |
| 4 | $N_1$ | 437 | 431 | 710 |
| | $N_2$ | 63 | 412 | 810 |
| 5 | $N_1$ | 445 | 535 | 690 |
| | $N_2$ | 410 | 549 | 823 |
| 6 | $N_1$ | 423 | 531 | 824 |
| | $N_2$ | 403 | 532 | 636 |
| 7 | $N_1$ | 406 | 513 | 692 |
| | $N_2$ | 334 | 528 | 712 |
| 8 | $N_1$ | 417 | 519 | 651 |
| | $N_2$ | 419 | 517 | 420 |
| 9 | $N_1$ | 428 | 516 | 882 |
| | $N_2$ | 430 | 538 | 797 |
| 10 | $N_1$ | 344 | 456 | 620 |
| | $N_2$ | 250 | 526 | 739 |
| 11 | $N_1$ | 304 | 549 | 699 |
| | $N_2$ | 99 | 384 | 402 |
| 12 | $N_1$ | 390 | 469 | 678 |
| | $N_2$ | 0 | 507 | 828 |
| 13 | $N_1$ | 298 | 468 | 356 |
| | $N_2$ | 84 | 540 | 846 |
| 14 | $N_1$ | 420 | 532 | 811 |
| | $N_2$ | 212 | 534 | 816 |
| 15 | $N_1$ | 418 | 550 | 860 |
| | $N_2$ | 339 | 508 | 872 |
| 16 | $N_1$ | 395 | 515 | 725 |
| | $N_2$ | 407 | 530 | 708 |
| 17 | $N_1$ | 440 | 490 | 818 |
| | $N_2$ | 279 | 553 | 850 |
| 18 | $N_1$ | 462 | 552 | 808 |
| | $N_2$ | 244 | 542 | 857 |

In this welding case study, the variation of static welding strength may increase very significantly if we change the welding conditions slightly. In other words, if we apply the traditional mean-value-only engineering approach and use the equivalent spot-welding length as the measurement characteristic, we may make a very significant misjudgment by ignoring variation effects. The author doubts the repeatability of this measurement characteristic (i.e., equivalent spot-welding length), since the welding engineers usually obtained very different results even when the settings of the welding machine remained unchanged. If variation in the objective measurement characteristic is great, further validation tests are required to estimate the mean value and the range (or distribution) of the collected experimental data, necessitating increased time, human resources, and money. The efficiency of the total-development process will consequently decrease. In the traditional product-oriented development process, tuning methods are very commonly applied to calibrate objective engineering characteristics to meet predetermined requirements. However, tuning methods are insufficient to meet downstream requirements if the engineering characteristics have too much variation. Take the data of $N_2$ of Experiment 8 of Table 4.1, for example. Assume that a static welding strength (i.e., 517 kgf) of a 20mm weld does not meet the strength requirement. Welding engineers will intuitively increase the welding (say to 30 mm) to reinforce the weld. However, if there is too much variation, this new welding length may provide even less static welding strength (only 420 kgf at 30 mm, Table 4.1). This kind of variation effect, in fact, occurs very frequently in industrial applications. In the experience of the author, most engineering problems that do not meet the predetermined requirements are caused by this type of variation effect.

Many design engineers think that if they understand all the design theories behind an engineering system and then apply these theories to design the system, the system will not have any significant quality problems. In fact, many quality problems can occur even when all design theories seem reasonable. For example, many engineers think that if they increase welding length, total welding strength should increase accordingly. This type of engineering thinking is based on mean-value-only theories, which assume no variation. In the real world, actual responses of engineering systems often differ from the responses predicted through design theories. On the other hand, many designs (e.g., universal joints) work very well, although theoretically they should not. In sum, at the new-technology-development stage, we need to make the real functions of the target technology as close as possible to ideal under noise conditions. After the technology is made robust against noise factors, it can be used to develop a family of future products. A robust technology is predictably flexible and provides a wide latitude for various design changes and operating conditions. To enhance total-product-development efficiency, we need to develop robust technologies even before new products are planned or developed. In this welding case study, the linearity and robustness (i.e., minimum variation) should be enhanced at the technology-development stage.

Although the author often chose to omit analytical details pertaining to case studies in order to focus the reader on basic functions and variations in engineering systems (and away from a traditional mean-value-only approach), a few comments on analytical methods in robust-technology development are pertinent. Experimental data illustrating the relationship between welding length and strength are shown in Fig. 4.7. Originally, the engineers applied ordinary linear-regression methods to analyze this data, without forcing the regression line through the zero point. However, such regression analyses are not realistic because there should be no welding strength if the welding length is zero. When unaware of the basic functions of their engineering systems, engineers may apply inappropriate analytical methods such as regression analyses or response surface methods. These analyses are meaningless from the viewpoint of robust engineering. As mentioned above, if welding length is zero, welding strength should be zero. We need to force the regression line through the zero point.

The most important issue in robust-technology development is to determine the ideal function for the targeted new technology. It is meaningless to randomly select specific regression forms (e.g., exponential equations) for regression analyses. Ideal function should be decided before regression analysis. The author derived the ideal function for laser-welding technology based on the following observations. If welding length is zero, welding strength should be zero. The welding strength should also be linearly proportional to the welding length.

In this case study, the ideal function of thin-plate (less than 1-mm thickness) welding indicates that static welding strength should be proportional to the square root of welding length. This is based on welding theories and engineering experience. Thus, we conducted a regression analysis

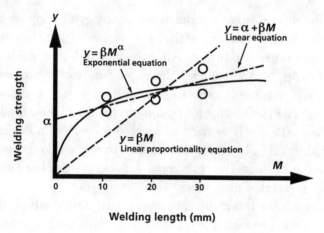

**Figure 4.7** Ideal function between welding strength and welding length through zero point.

(through the zero point) between welding strength and the square root of welding length, as shown in Fig. 4.8, then applied the results of the regression analysis and dynamic type S/N ratio to evaluate the functional robustness of thin-plate welding. After determining the optimal conditions, we validated the results using optimal settings versus original settings.

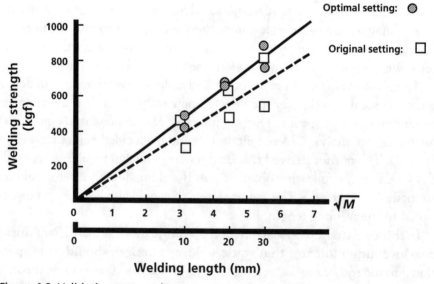

**Figure 4.8** Validation test results.

Validation test results are shown in Fig. 4.8, where we can see that welding strength increased significantly and functional variation decreased under optimal conditions. In the case of a 20-mm weld, welding strength at optimal settings increased by 13% and variation was reduced to one-fifth of original settings. The major problems in laser-welding technologies were thus solved, and the newly developed technology readily replaced conventional spot-welding technologies.

After the laser-welding technology was made robust, it was applied to numerable products. One engineer stated:

> I am not directly involved in laser-welding technology development. I was, however, involved in the introduction of this laser-welding machine at our company and was disappointed to see it sit idle and unused in production processes for 1-1/2 years due to numerous unsolvable problems. I am very happy to see that this machine can now be used in the mass-production process.

The author was able to help welding engineers apply this technology in welding high-pressure containers, a process in which welding strength is critical. Traditionally, the number of blowholes was used as the measurement characteristic for welding strength. This was an extremely inefficient measurement characteristic because quality inspectors used subjective descriptions and needed to inspect all welding lines to determine defects. If one welding line had more than a certain number of blowholes, it would be judged defective. Although engineers tried to improve this process for more than 1-1/2 years, failure rates remained around 40 to 50%. When they asked for help, the author immediately told them that the number of blowholes was a very downstream quality characteristic and not an efficient measurement characteristic for this welding process.

For the development of new technologies, rather than using downstream appearance characteristics, such as blowholes, we should apply upstream measurement characteristics, such as welding homogeneity, to measure the functional robustness. As mentioned above, the failure rates of the welding process of high-pressure containers were around 40 to 50%. In other words, about half of the welding lines had blowholes, which occurred due to nonhomogeneity of the welding power supply. This meant that the laser-welding machine was not supplying very homogeneous thermal welding power on the welding lines of high-pressure containers. Thus, if the welding lines were very long, the possibility of blowholes was high. To make this welding process robust, we first needed to make the welding machine provide homogeneous welding energy on welding lines of variable lengths. If this objective were achieved, there would be no blowholes. The measurement of the nonhomogeneity of the welding power supply is the functional variation shown in Figs. 4.9 and 4.6. The only difference between these two figures is that the welding length settings of Fig. 4.9 are much greater than those of Fig. 4.6. We expected the thermal energy per unit of welding length to be close to constant. Thus, the variation of the proportionality constant between welding strength and welding length could be applied as the measurement of the homogeneity of the welding power supply.

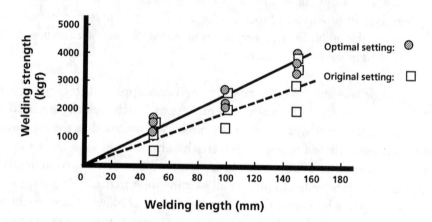

**Figure 4.9** Validation test results for welding of high-pressure containers.

The plates (over 4 mm) used in the welding process in Fig. 4.9 are much thicker than those (less than 1 mm) in the case study in Fig. 4.8, where the ideal function was to have welding strength be proportional to the square root of welding length. However, the ideal function of the case shown in Fig. 4.9 is that the static welding length be proportional to corresponding welding length. Because of different ideal functions, the energy supplies per unit of welding length of these two cases are very different. We conducted the same experiments as illustrated in Table 4.1, and the validation results of optimal settings versus original settings are illustrated in Fig. 4.9. Distributions of welding strength in optimal and original settings are also illustrated in Fig. 4.10. The process-capability (PC) value of the welding shear strength increased from 1.6 to 2.7, primarily due to the improvement of welding power supply homogeneity. Lastly, the engineers conducted validation tests by welding 150 high-pressure-class plates and checked their blowholes. There were no blowholes. It took only two months to make this laser-welding technology robust and to apply it to mass-production processes. Before this, the laser-welding machine had sat idle in the plant for 1-1/2 years and the welding engineers had spent a tremendous amount of energy trying to improve the machine without any significant results.

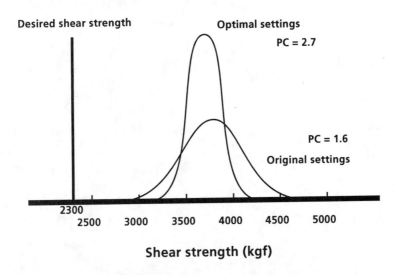

**Figure 4.10** The welding-strength distribution of products. (PC stands for process-capability.)

By including this welding case study, the author wanted to emphasize the importance of improving the basic function of an engineering system instead of measuring downstream quality characteristics. Only through the improvement of basic functions can we develop robust technologies. Robust-technology development is an extremely powerful way to improve the productivity of a total-product-development process. It is a critical to shift the paradigms of engineers from a product-oriented approach to a technology-oriented approach. Only then can we really restructure the technological bases of manufacturing companies.

# CHAPTER 5

# The Concepts of Parameter Design (Robust Engineering)

## 5.1   ROBUSTNESS AGAINST NOISE FACTORS

The most important issue in the product-development process is to make target products (or processes) robust against both environmental and deterioration noise factors. Every stage in the product-development process is critical to the achievement of this objective; however, the most cost and time efficient way to achieve it is at the research and development stage.

When technologies developed slowly, it was possible to apply the design-test-fix type of development processes to improve product quality through the evolutionary process. High-quality products could evolve slowly because technologies developed slowly. However, current technological development is so fast that the design-test-fix type of development processes can no longer ensure the quality of new technologies.

Many new technologies are developed from new engineering fields in which little knowledge or experience has been accumulated. There are no lessons, no reference manuals for the development of these new technologies. It can be difficult to determine the measurement techniques that would be appropriate for evaluating the performance and quality of the new technologies. Even when appropriate measurement techniques exist, they may not be sufficient to improve the robustness of new technologies in a cost and time efficient manner. Because new-product development is now much more complicated and sophisticated, conventional quality-engineering techniques are not able to ensure the quality of newly developed technologies. If development engineers continue to apply conventional quality-engineering methods in industrial research and development, they will not be able to develop robust technologies in a timely manner. Many key quality characteristics may be overlooked because of the increasing sophistication and complexity of new technologies. Increased efficiency in the total development process is critical to the competitiveness of any manufacturing company.

Engineers work hard to solve the quality problems they encounter. Unfortunately, they usually apply scientific-type research activities to determine the possible root-causes of their downstream quality problems and then try to apply fire-fighting types of activities to eliminate them. Solving quality problems in this fashion will render engineering systems more complicated than necessary and may result in many negative side effects.

In the experience of the author, the most important issue in the product-development process is to apply robust engineering methodologies to desensitize target engineering systems against noise factors, not to eliminate all root causes. In other words, development engineers need to make the basic functions of their target engineering systems insensitive to downstream noise factors.

Such quality-engineering methods, initiated and promoted by Dr. Genichi Taguchi, are called parameter-design (e.g., robust engineering) methods. Applying conventional quality-engineering methods to eliminate noise factors takes much time and engineering resources, does not guarantee results, and may leave noise factors that can still impact the target engineering systems. It is much easier to make the engineering system robust against noise factors. Through robust engineering methods, the effects of noise factors will become negligible.

## 5.2 THE SHIFT FROM PRODUCT-ORIENTED DEVELOPMENT TO TECHNOLOGY-ORIENTED DEVELOPMENT

Conventional quality-engineering methods focus on how to make the downstream characteristics of target products meet the requirements and needs of customers. Those development processes that focus on specific customer needs and requirements are called product-oriented development processes. However, in quality engineering, it is more cost and time efficient to improve and enhance the robustness of the basic functions of the target engineering systems before downstream customer needs are identified and specified. After the functional robustness of the target system is enhanced, downstream quality characteristics can be adjusted to meet various customer requirements at later stages in the development process. As opposed to the product-oriented development process, this type of development process is called the technology-oriented development process.

Currently, most Japanese manufacturing companies apply product-oriented development methods, in which downstream customer requirements are identified before all other development activities are embarked upon. These development processes therefore focus on how to adjust the mean values of the target quality characteristics to meet various requirements. Quality-engineering activities cannot be conducted before downstream customer requirements are identified. Development activities can only be conducted after product planning. Product-oriented development processes are usually associated with the time-consuming design-test-fix type of quality-engineering methods, in which development engineers attempt to determine the root causes of downstream quality problems. (Unfortunately, even if development engineers can determine the root causes, they may be unable to determine appropriate solutions.) For instance, in new-automobile development, most companies develop automobile prototypes (or component prototypes) first and then test these prototypes under various environmental conditions to determine all possible downstream quality problems. Next, they try to fix the identified problems by redesigning. In other words, actual quality problems must first occur in order to identify and solve the quality problems. Thus, product-oriented development processes associated with design-test-fix quality-engineering methods are very time inefficient. It is very difficult to use a limited number of prototypes to simulate all possible operating conditions of mass-produced products. Important downstream quality characteristics may be overlooked, allowing severe malfunction problems to occur under customer-usage conditions and causing customer dissatisfaction.

One critical issue in the technology-oriented development process is to predict critical downstream noise factors that may deteriorate the basic function of the target engineering system and to apply robust engineering methods to desensitize the system to these noise factors. The most efficient way to do this is to reduce the functional variation of the system through robust engineering methods. Currently, no other quality-engineering methods have proven as efficient in reducing functional variation; this is because the variation effects of downstream noise factors on basic functions can be reduced at a very early stage in the product-development process. Robust engineering activities can prevent serious downstream quality problems, and thus allow products to be developed in a timely manner.

In the product-oriented development process, development engineers usually focus on adjusting target quality characteristics to meet customer requirements, but they seldom consider reducing functional variation in the engineering systems of the product. As a result, the original quality of the target engineering system is not enhanced. Because the basic functions of the target systems vary tremendously, some serious downstream quality problems may occur.

To solve downstream problems efficiently, development engineers must focus their quality-engineering resources on improving the robustness of basic functions at a very early stage of their product-development processes. In the product-oriented development process, product-planning and marketing research are conducted before any other product-development activities. In the technology-oriented development process, functional robustness is enhanced before any product planning or marketing activities occur and target quality characteristics are adjusted to meet customer needs and requirements at even later stages. Through technology-oriented development processes, high-quality products can be developed to meet ever-changing customer requirements in a time-efficient way.

## 5.3   PARAMETER DESIGN (ROBUST ENGINEERING)

In the product-oriented development process, development engineers try to determine the target downstream quality requirement (e.g., $A_0$ in Fig. 5.1) first and then adjust the control factor ($a_0$—or $b_0$, $c_0$, if more than one control factor) so that the objective quality characteristic ($A$) will meet its target ($A_0$). There may be infinite sets of control factors that can make $A$ meet its target $A_0$. In general, we only need one set of all possible solutions to meet the requirement. Unfortunately, most engineers still spend excessive engineering resources in adjusting the mean value of $A$ to meet the target requirement $A_0$, instead of reducing the variation of $A$. In fact, the variation of $A$ is a measurement of the robustness of $A$. In a robust engineering procedure, the major concern is on how to reduce the variation around the objective characteristic(s). Because of numerous downstream noise factors, the target quality characteristics will vary somewhat around their target values. Reducing this variation is the most critical issue of the robust engineering procedure. Conversely, the major focus in conventional quality engineering is on tightening component and material tolerances to reduce the transmitted variation of the assembled products. Thus, the percentage of defective components or materials produced through conventional quality engineering methods will be very high due to the strict tolerance requirements. In addition, the manufacturing cost achieved through conventional quality engineering methods is usually higher than those through robust engineering methods because of the increased adjustments and scraps generated by conventional quality engineering methods. The author refers to those development processes that use conventional quality engineering as "tuning-type" quality engineering methods (i.e., the product-oriented development process).

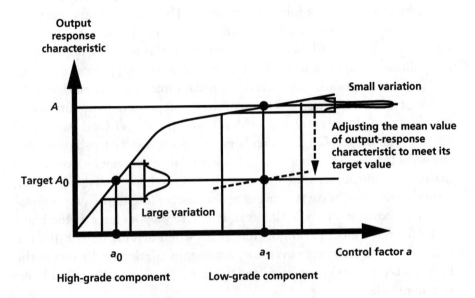

**Figure 5.1** The concept of robust engineering. Shown is a two-step optimization method to achieve robustness: (1) The first step is to reduce the variation of the basic function by a good combination of control factor(s). (2) The second step is to calibrate adjustment factor(s) so that output-response characteristics meet target values.

One efficient way to solve downstream quality problems is to determine the optimal settings of controllable factors (or design parameters) so as to minimize the functional variation of the target systems by rendering the target response characteristic(s) insensitive to downstream noise factors. The next step is to calibrate adjustment factors to meet their downstream requirements. Through robust engineering methods, development engineers can develop high-quality products that are robust under various downstream usage conditions.

## 5.4 USING THE NONLINEARITY EFFECTS BETWEEN OUTPUT RESPONSES AND CONTROL FACTORS TO ACHIEVE ROBUSTNESS

In this section, the example of an automobile brake system will be used to illustrate the concept of robust engineering. The objective function of an automobile brake system is to stop the car. Generally speaking, an ordinary driver expects the braking force (output) of an automobile brake system to be linearly proportional to the force that the driver applies to the brake pedal. If the relationship between the pedal force (input signal) and the braking force (output response) is not linearly proportional, the automobile may be stopped violently and suddenly, and the driver (and passenger) thrown from the car. On the other hand, if there is no relationship between the pedal force and braking force, the automobile would not be stopped, no matter how much force the driver applies to the brake pedal. Generally, most drivers want brake systems to perform according to the proportional function shown in Fig. 5.2. However, if a driver applies force to the brake pedal for too long a period of time (e.g., when driving downhill), the brake shoes (or pads) may overheat, causing the friction coefficient of the brake pads (or shoes) to decrease and the linear proportionality of the brake system to fade.

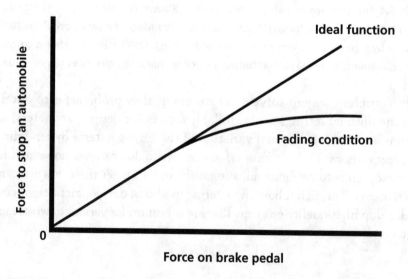

Figure 5.2 Brake-system function.

The curve shown in Fig. 5.2 illustrates fading in a brake system, where the linear proportionality of the brake system deteriorates and nonlinear effects become significant. In many engineering systems, input signals and output responses are expected to be linearly proportional, and so we need to ensure the linear proportionality of the basic functions of the target engineering systems under all environmental and usage conditions. This is the key issue in robust-technology development.

**TABLE 5.1  An $L_{18}$ orthogonal array**

| Factor No. | A | B | C | D | E | F | G | H | S/N$\eta$ (dB) | Sensitivity S (dB) |
|---|---|---|---|---|---|---|---|---|---|---|
| 1 | 1 | 1 | 1 | 1 | 1 | 1 | 1 | 1 | 31.41 | -0.0022 |
| 2 | 1 | 1 | 2 | 2 | 2 | 2 | 2 | 2 | 39.70 | 0.0058 |
| 3 | 1 | 1 | 3 | 3 | 3 | 3 | 3 | 3 | 39.68 | 0.0028 |
| 4 | 1 | 2 | 1 | 1 | 2 | 2 | 3 | 3 | 9.25 | 0.0730 |
| 5 | 1 | 2 | 2 | 2 | 3 | 3 | 1 | 1 | 44.56 | -0.0001 |
| 6 | 1 | 2 | 3 | 3 | 1 | 1 | 2 | 2 | 42.02 | 0.0020 |
| 7 | 1 | 3 | 1 | 2 | 1 | 3 | 2 | 3 | 33.75 | 0.0057 |
| 8 | 1 | 3 | 2 | 3 | 2 | 1 | 3 | 1 | 44.59 | 0.0003 |
| 9 | 1 | 3 | 3 | 1 | 3 | 2 | 1 | 2 | 19.18 | 0.0114 |
| 10 | 2 | 1 | 1 | 3 | 3 | 2 | 2 | 1 | 42.80 | 0.0011 |
| 11 | 2 | 1 | 2 | 1 | 1 | 3 | 3 | 2 | 30.55 | 0.0145 |
| 12 | 2 | 1 | 3 | 2 | 2 | 1 | 1 | 3 | 26.41 | 0.0166 |
| 13 | 2 | 2 | 1 | 2 | 3 | 1 | 3 | 2 | 25.36 | 0.0148 |
| 14 | 2 | 2 | 2 | 3 | 1 | 2 | 1 | 3 | 35.24 | 0.0056 |
| 15 | 2 | 2 | 3 | 1 | 2 | 3 | 2 | 1 | 42.52 | 0.0022 |
| 16 | 2 | 3 | 1 | 3 | 2 | 3 | 1 | 2 | 41.01 | -0.0009 |
| 17 | 2 | 3 | 2 | 1 | 3 | 1 | 2 | 3 | 2.63 | 0.1801 |
| 18 | 2 | 3 | 3 | 2 | 1 | 2 | 3 | 1 | 39.30 | 0.0025 |

Again, the first step in robust engineering is to reduce the functional variation of the target engineering systems, after which the mean values of the objective quality characteristics are adjusted to meet the requirements of customers.

In typical robust engineering projects, there may be more than 20 control factors that significantly affect the objective quality characteristics. (Design parameters are usually called control factors in robust engineering.) Among all possible control factors, we select 8 of the most promising on which to conduct experimental robust engineering. We can assign these 8 factors to an orthogonal array, for example, an $L_{18}$ orthogonal array (Table 5.1). Next, empirical experiments or computer simulations are conducted to collect response data for the objective output responses. An $L_{18}$ orthogonal array is highly practical in this regard, and at most, 8 control factors (design parameters) can be included. Among the 8 control factors tested, we usually find two or three that are especially significant in affecting the variation of the basic function of the target engineering system.

The first step in robust engineering is to apply experimental design methods to enhance the stability (i.e., robustness) of the basic function of the engineering system; the second step is to calibrate adjustment factors to adjust the mean value of the objective response characteristic(s) to meet the needs of customers. Adjustment factors are those design parameters (control factors) that do not affect significantly the robustness of the basic function, but that do significantly affect the mean value of the objective output-response characteristic. If we have 8 control factors, we can usually isolate one or two adjustment factors. Generally, only one adjustment factor is needed to calibrate the mean value of the objective output-response characteristic(s).

An example involving quartz (or ceramic) vibrators in analog watches will be used to illustrate the two-step optimization of robust engineering methods (Fig. 5.3). The objective function of a quartz (or ceramic) vibrator is to accurately regulate the expressed time of the watch to be close to the real time. To achieve this objective, we need to ensure that vibrator frequency remains stable (i.e., shows little variation) under all kinds of environmental and usage conditions.

Vibrator frequency variation will affect the accuracy of the watch. We can measure all kinds of vibrators to determine which is the most stable (i.e., has minimum frequency variation). Assume that the selected vibrator has a frequency of 200 hertz. We can set the gear ratio of the watch to be 1 second per 200 vibrations. Therefore, the first step in the robust engineering is to reduce frequency vibration, and the second step is to adjust the gear ratio of the watch to make the expressed time match real time. In this vibrator example, gear ratio is an adjustment factor for the quartz (or ceramic) watch.

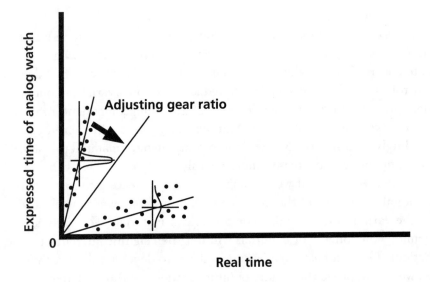

**Figure 5.3** Robust engineering for quartz (or ceramic) watch vibrators. The two-step parameter-design procedure for quartz (or ceramic) watch vibrators is: (1) Try to find a vibrator that is stable and of little variation. Note: It is unnecessary to adjust the expressed time of the watch to match real time at this step. (2) Apply an adjustment factor (e.g., gear ratio) to make the expressed time of the watch match real time.

Using this approach, we can produce very accurate watches at low cost by using a low-cost (but robust) ceramic vibrator instead of an expensive quartz vibrator. Again, the most important issue in the robust-design procedure is how to enhance the functional robustness of the target engineering system. The most critical issue is not whether expressed time is close to real time.

To enhance the functional robustness of a target engineering system is the same as to improve the stability of its basic function. We need to minimize the system's functional variation to ensure functional stability. Functional stability (i.e., robustness) needs to be ensured even before product planning. The objective quality characteristics can be adjusted to meet target values at later stages of the product-development process through the calibration of adjustment factors. Using two-step robust engineering methods, high-quality products can be developed quickly. These upstream quality-engineering activities allow quality to be designed into the target engineering systems even before new products are planned and thus reduce development time.

Robust engineering is a very cost- and time-efficient way to reduce the functional variation of target engineering systems, but it is usually difficult to control functional variation in mass-produced products. If we do not apply robust engineering methods to reduce functional variation, our only other option is to tighten the tolerances of incoming components or materials, an expensive proposition. Manufacturing cost will then increase accordingly. In other words, if we loosen component or material tolerances, product quality will decrease simultaneously.

On the other hand, if we can apply robust engineering to improve the functional robustness of the target engineering system, we can desensitize it to the variation in incoming components or materials. The variation in incoming components or materials will not then significantly affect basic functions. Using robust engineering methods to develop robust technologies, we can improve the quality of future products without tightening the tolerances of key components or materials and can still apply low-cost materials or components to make high-quality products. Through the implementation of robust engineering methodologies at a very early stage in the product-development process, we can make high-quality products while simultaneously reducing cost.

The two objectives of robust engineering are:

- To improve the functional robustness of the target engineering system, after which objective quality characteristic(s) can be quickly adjusted to meet various requirements.
- To ensure that the basic functions of the target system remain stable, without requiring compensation adjustments, throughout its operational life.

The most important issue in achieving these two objectives is not data analysis or experimental design, but learning how to formulate the robust engineering project and how to select the basic function and measurement characteristics of the target systems. It is critical to determine the functional limits of the target system at a very early stage in the product-development process. Development engineers should be shown how the functional robustness of the target engineering system affects its durability. As functional robustness should exceed customer requirements, development engineers should devote themselves to enhance it. If development engineers are not able to determine the functional limits of their engineering systems, they will not be able to ensure product robustness under customer-usage conditions, regardless of the amount of money and engineering sources used in product development. Development engineers need to develop highly robust and flexible systems in a very early stage of product development, which in turn will enable them to develop in a timely manner high-quality products that meet market requirements.

The advantages of robust engineering methodologies are:

- Variations in incoming materials and components will entail variation in basic functions of the target engineering systems. Using robust engineering methods, we can improve the functional robustness of the target engineering system in a time- and cost-efficient manner.
- Under customer-usage conditions, the quality characteristics of components or materials may deteriorate. However, the basic functions of the target systems will be unaffected if functional robustness has been enhanced.
- Using robust engineering, we can even use low-grade components or materials (e.g., $a_1$ of Fig. 5.1) to make high-quality (i.e., low variation) products, if the basic functions of the engineering systems of the products have been made robust. Thus, we can achieve high-quality while reducing cost.
- Similarly, process engineers can loosen the tolerances of key components or materials in order to increase manufacturing speeds and productivity, while simultaneously reducing cost.

In conclusion, development engineers need to apply robust engineering methodologies to ensure the functional robustness of their target engineering systems at a very early stage in the product-development process. Robust engineering methods can make this process more time and cost efficient than is possible under conventional quality engineering methods, which are based on the assumption that quality is inversely proportioal to cost. (Traditionally, we assume that if we want to improve the quality of products, we need to tighten the tolerance specifications of key quality characteristics, which usually increases manufacturing cost.) However, if we can apply robust engineering methods to develop robust and flexible product or process technologies at very early stages in the total-product-development process, we will be able to improve quality while simultaneously reducing cost. If the sensitivity between target output response and input signal is very high, we may also reduce the overall size (and thus weight, energy consumption, etc.) of the target engineering system while maintaining functional robustness. Eventually, customer satisfaction will improve as well.

If we can apply robust engineering methodologies to enhance the functional robustness of target engineering systems at a very early stage in product development, quality and cost need not oppose each other. To survive in this highly competitive market, we need to develop high-quality products at low cost, and to deliver them to market before the competition. To achieve this objective, we need to bring robust engineering upstream to the technological research and development stages so that we can develop robust product and process technologies.

# CHAPTER 6

# Company-Wide Implementations of Robust Engineering

## 6.1   BACKGROUND

As mentioned in Chapter 1, top management in automotive companies often refers to the *Maritz National Car Quality Tracking Study* or J.D. Power reports to evaluate performance (or deterioration) of its products under customer-usage conditions. Generally speaking, if the initial quality of a mass-produced product has appreciable variation, it is likely to deteriorate significantly under customer-usage conditions. According to these reports, the initial quality of automobiles produced by Toyota Motor Co. is usually of less variation than those of Nissan manufacture. These reports also indicate that the deterioration rates of Toyota models are usually less than Nissan's. This significant quality difference between automobiles produced by these two companies led the author to suspect that Toyota Motor Co. and Nissan Motor Co. might employ different quality-engineering methodologies. Extensive investigation of the development processes at Toyota Motor Co. led the author to realize that Toyota had already popularized robust engineering methods throughout the company 30 years ago and had also incorporated robust engineering into its product-development processes. He also discovered that robust engineering has become the daily norm for Toyota engineers. Since robustness has been intensively designed into Toyota automobiles, they are likely to be more durable and reliable than those manufactured by other companies. The author believes that robust engineering may be the primary reason that automobiles produced by Toyota Motor Co. show less variation and deterioration than those produced by other companies.

Armed with this information, the author convinced top management at Nissan Motor Co. to introduce robust engineering. However, no one in the company really knew how to popularize robust engineering company-wide or how to incorporate robust engineering into the product-development processes. It was a further challenge to shift the thinking processes of engineers and management personnel from traditional quality-engineering paradigms (i.e., solving existing problems) to robust engineering paradigms (desensitizing against downstream noises). Some individuals still stuck to technologies that had been proven to be unreliable or noncompetitive. Many in management (especially middle to upper level) felt that they were already successful in their personal careers and were not willing to invest engineering resources in developing new technologies to enhance the competitiveness of the company. Many others were comfortable with the slow advancement provided by already outdated mechanical technologies. Such resistance did not diminish the need for robust engineering, however. While current automotive components are primarily based on electronic/semiconductor technologies that develop very fast, management usually increases engineering resources and testing equipment based on the slow growth rate of traditional mechanical technologies. For example, the increase in the number of engineers and testing equipment was less than twofold during the last decade, insufficient to catch up with advances in electronic/semiconductor technologies, causing many engineers to work frequently until midnight in an attempt to keep pace with development schedules. Unfortunately, this company could not really afford to increase engineering resources and testing equipment in proportion to the fast advancement rate of new electronic/semiconductor technologies. Obviously, more efficient engineering methods for the development of new products or technologies needed to be introduced.

Not many people really understood how important it was to improve the efficiency of the technology-development process. Quality-engineering activities at the company were primarily conducted to reduce downstream warranty cost or customer dissatisfaction. To solve quality problems efficiently and fundamentally, the company would have to bring its quality-engineering resources upstream so as to develop competitive and robust technologies. In the opinion of the author, the most efficient way to enhance product quality is to develop robust technologies and use them to develop future products and manufacturing processes that will remain reliable and predictable under various operating or usage conditions. In this way, many downstream quality-engineering activities, such as on-line inspection, after-market services, warranty, etc., could be minimized or even eliminated. Since not many engineers at that time were familiar with the methods and theories of robust engineering, the author decided to introduce and popularize robust engineering at the company.

## 6.2 INTRODUCING ROBUST ENGINEERING INTO THE COMPANY

In the beginning, only a few people volunteered to help with the introduction of robust engineering. A focus team was formed to initiate introduction and popularization of robust engineering in July 1988, and a robust engineering department was scheduled to be set up in January 1989. The author wrote a proposal for the objective, mission, and organization of the department, which was approved by top management in late 1988.

The robust engineering department was formally named the Reliability Engineering Center. The first step was to choose its location. The department was finally located in the technical center (instead of the proving ground) because robust engineering is supposed to be incorporated into the upstream engineering activities (e.g., development, research, design) performed at the technical center, instead of the downstream validation tests done at the proving ground.

The author was appointed as manager of the department and was responsible for the introduction and popularization of robust engineering. Six months after the setup of the department, five new employees were formally assigned to it and the volunteers went back to their original assignments. Promotion activities for robust engineering formally commenced in July 1989.

At the introductory stage, the mission of the department was "to improve product quality while simultaneously reducing the associated manufacturing cost." Currently, the mission of the department has been updated to read: "to develop generic and robust technologies to make the company technologically competitive and to reduce product-development time." Simply put, the current mission of the department is to change the development processes of the company from being specific-products oriented to generic-technologies oriented. The goal of the department's 10-year plan is for all engineers and management in the research, development, and design departments to use robust engineering as their primary tool for the development of generic and robust technologies within the 10-year period. To achieve this goal, the author expects that some of those engineers who believe in robust engineering will be promoted to top management within 10 years. There are about 2000 engineers and managers in the research, development, and design departments. Assuming that 6 to 7 engineers or managers are involved in one robust engineering project, through simple calculation, at least 300 robust engineering projects each year need to be conducted for all engineers (except technicians) and managers in these departments to be involved in the implementation of robust engineering.

Several approaches were used to promote and popularize robust engineering methods throughout the company. The first was to conduct case-study workshops under the guidance of Dr. Genichi Taguchi. Because of Dr. Taguchi's profound knowledge and industrial experience in robust engineering, many engineers and managers were willing to attend the workshops. The most important issue in the popularization of robust engineering is to use on-the-job training to teach engineers how to initiate and formulate robust engineering projects. Through on-the-job training, team members in robust engineering projects can share a common learning environment and training material. This approach helps both engineers and management understand the basics of robust engineering. It is not time efficient to ask engineers to take outside college courses on robust engineering individually. Consequently, the engineers at Nissan were encouraged to use on-the-job training to become familiar with industrial robust engineering applications (instead of academic theory) and then use what they had learned in their daily assignments. Listed below are the seminars and training workshops provided by the robust engineering department for engineers and management.

1. Robust engineering seminars and workshops
   (instructed by the author and colleagues):
   a. Introduction seminars (2 days): 4 seminars per year,
      35 to 50 people per seminar
      • Concepts of robust engineering (first day)
      • Basic analytical methods of robust engineering (second day)
   b. Training workshops (3 days): 4 workshops per year,
      30 people per workshop
      • Dynamic-type robust engineering (1 day)
      • Experimental-design example using paper helicopters and
        calculation of total quality loss (5 teams, 2 days)
   c. Class for new management (1 day): 1 class per year
      • Concepts of robust engineering

2. Case-study workshops
   a. Case-study workshops instructed by Dr. Genichi Taguchi:
      10 to 15 workshops per year
   b. Case-study workshops instructed by the author:
      available on request

Illustrating successful case studies proved to be a very convincing way to popularize robust engineering methods throughout the whole company. One engineer from each research and design department was designated as the resident expert to lead the first robust engineering project in the department. In order to integrate robust engineering with other engineering activities, such leading engineers need to have profound knowledge of and experience with the target engineering system and its integration with other systems. Take the development of a new vehicle, for example. A vehicle's suspension system must be integrated with other engineering systems, such as the chassis, steering, etc. Merely improving the suspension system cannot improve the ride of the vehicle. To ensure the total quality of the whole car, the target system (e.g., suspension) needs to be integrated with such other systems as engine, transmission, exhaust system, chassis, and so on. It is vital to apply robust engineering to product (e.g., vehicle) systems to ensure the total functional robustness of the product. If any system fails to perform its basic function, the product is considered nonrobust.

At the introductory stage, the focus is on how to popularize the concepts and methods of robust engineering in every department. Popularizing robust engineering concepts and methodologies in every corner of the company is critical to the company-wide implementation of robust engineering. These concepts and methodologies are very different from those in traditional quality-engineering methods, and therefore, it is not always easy for all engineers to accept them. Some people, especially those who habitually resist new methodologies, are likely to treat robust engineering as another new, but unproved, engineering method until convinced of its value and effectiveness. These people usually like to stay with traditional engineering methods, and may be against robust engineering, even going so far as to try every means to resist its popularization. Fortunately, some people are more open-minded and believe that robust engineering is a valuable tool to be used in conducting engineering activities. It was such individuals who generally tried their best to help the author's department popularize robust engineering methods, and they took every opportunity to illustrate the importance and necessity of the implementation of robust engineering in their departments.

To popularize robust engineering throughout the company, at least one or two supporters are needed to promote robust engineering activities in each department (especially in the research, development, and design departments), preferably individuals who have influence on the whole department, such as managers or senior engineers. There obviously will be conflict between adherents and nonadherents during the popularization of robust engineering methods, but the following suggestions may help resolve these conflicts and popularize robust engineering methods in an efficient way.

Most people simply treat robust engineering as a new engineering tool, but if the effectiveness and power of robust engineering methods can be demonstrated using real-life case studies (especially those with visible hardware), engineers may come to believe in these methods. Therefore, one efficient way to popularize robust engineering methods is to use— particularly in introductory training courses— successful case studies (especially from within the company) to demonstrate the effectiveness and power of robust engineering methods. In this way, more and more people may be convinced of the value of robust engineering. If people believe in robust engineering methods, they will try every means (e.g., word of mouth, implementation, etc.) to help with their popularization.

In the experience of the author, it takes time to make engineers and management truly understand the concepts and methods of robust engineering and to popularize robust engineering throughout the company.

In order to help engineers implement robust engineering methods, we need to provide on-the-job training in robust engineering. These methods are very different from traditional quality-engineering methods (e.g., statistical process control, fault-tree analysis, and so on); so in introductory training, instructors should clearly contrast the fundamental concepts of robust engineering methods to those of traditional quality-engineering methods. The most convincing way for engineers and managers to understand the methodologies is to help them implement robust engineering methods on real-life projects.

Engineers and managers can learn robust engineering methods individually, but this is, unfortunately, not the most efficient way to popularize robust engineering methods throughout the company. It takes engineers and management more time and resources to learn robust engineering methods individually than as a team. It is more productive to provide on-the-job training to the whole department (or project team) than to individuals. At Nissan, after providing introductory training on robust engineering to each department, instructors held workshops to help the engineers conduct specific projects through customized consultations. The introductory course in robust engineering methods is a two-day course. The first day focuses on illustrating the fundamental concepts of robust engineering methods and the reasons for their use, as opposed to traditional quality-engineering methods. The second day focuses on very simple experimental-design and analysis techniques in robust engineering. Sophisticated technical details in robust engineering are not taught in this introductory course, since the objective of the introductory course is very different from the training of resident experts. In the former, only the fundamental concepts and simple techniques of robust engineering are illustrated, while in the latter, sophisticated technical details are taught.

## 6.3  WORKING ON CHALLENGING PROJECTS

Obviously, many are willing to attend a course or workshop on robust engineering, but they are more interested in enhancing their knowledge than implementing robust engineering in their daily routine. However, even at the introductory stage, we hoped that some real-life case studies could be successfully conducted. One efficient way to determine potential successful case studies is to consult with engineers or managers who have challenging projects that are difficult to conduct through traditional quality-engineering methods (e.g., statistical process control, fault-tree analysis, failure modes and effects analysis). When instructors held robust engineering workshops, they focused their resources on those engineers and managers who had very challenging projects. Because of the difficulties in conducting such projects, participants were willing to try robust engineering methodologies instead of traditional quality engineering methods. To enhance the implementation of robust engineering methods, some restrictions were set on courses and workshops. It is not reasonable to mandate that all engineers and managers attend robust engineering training, as it is too expensive and time consuming. The objective of company-wide implementation of robust engineering is to improve productivity and product quality, not really to ensure that everyone obtain a knowledge of robust engineering. In other words, it is wasteful to force everyone to undergo training in robust engineering methodologies. However, it is critical to ensure that all persons who need such training get it.

Some in top management indeed asked the author to provide training in robust engineering to every engineer and manager. Generally speaking, top management is seldom involved in such training and may be unaware that providing training in robust engineering methods to every engineer and manager will not guarantee that all participants will implement it in their daily routine. Many employees may attend training and then put the training material away in their cabinet instead of implementing it. The author insisted on providing training in robust engineering to those who really needed it (e.g., engineers and managers involved in challenging projects), not to everyone in the company.

Those selected for the training (or workshops) need to initiate and formulate their engineering problems beforehand. (At Nissan, they were required to adjust their project schedule for the robust engineering workshop.) They also need to check with the consultants to determine whether robust engineering methods can really solve their engineering problems. Generally speaking, those participants who have challenging problems are more enthusiastic and more willing to implement robust engineering in their projects than others.

In short, the content of robust engineering courses and workshops is not necessarily the most important issue. More important may be how to make such training and workshops available to the people who really need them.

At the project-initiation stage, many people tend to limit themselves to very simple and easy problems. They think that they know little about robust engineering and thus want to start simply, with projects that have only two or three factors. However, from the viewpoint of robust engineering, this thinking is somewhat inappropriate. In robust engineering, engineers take advantage of the interactions between control and noise factors so as to desensitize the target systems against the noise factors. If engineers work on simple and small projects, they may not have much opportunity to take advantage of the interactions between control and noise factors. As a result, the effectiveness of robust engineering methods may not become evident. To help engineers understand the true meaning and effectiveness of robust engineering methods, engineers were asked to select projects that have a reasonable (5 to 8) number of control factors. In this way, the engineers were likely to have more opportunity to take advantage of control-by-noise interactions to reduce the functional variation of the target engineering system. If engineers can thereby solve some challenging engineering problems (especially those that have remained unsolved for years), they will be more convinced of the power and effectiveness of robust engineering methods.

In some projects, engineers may not be able to get satisfying results even through the exhaustive implementation of robust engineering. One possible reason may stem from an inability to measure the basic function of the target engineering system.

Another important point is that robust engineering should be used to solve upstream problems in basic technologies and not attempt to eliminate downstream quality symptoms. To improve the functional robustness of basic technologies, it is necessary to conduct well-organized training courses and workshops that help engineers (especially those in research, development, and design departments) initiate and formulate robust engineering projects for technology development. Case-study workshops are especially useful as a means of providing consultations to those engineers who are conducting robust engineering projects. These workshops can be held at the following three stages: (1) project planning, (2) analysis of experimental data, and (3) validation tests. Of these three consultation stages, project planning is the most important. At the project-planning stage, robust engineering instructors need to illustrate clearly how to evaluate the robustness of the basic functions of the target engineering systems (or new technologies), as opposed to identifying the associated downstream quality characteristics of these basic functions. Robust engineering is very different from traditional design-test-fix type of engineering. At the planning stage of any project, engineers should avoid any design-test-fix activities. In the opinion of the author, about 80 to 90% of the consultation effort at the planning stage should be focused on the selection of the basic functions of the target engineering systems (or technologies) and on how to measure the functional robustness of the systems.

## 6.4 CONSULTING ON ROBUST ENGINEERING PROJECTS AND TOP-DOWN POPULARIZATION

Engineers can apply either computer simulations or empirical experiments to evaluate the functional robustness of the target systems (or technologies), based on considerations of scope, cost, and hardware feasibility. Collected experimental data and analytical results need to be discussed by everyone on the project team to determine the optimal settings for control factors, so as to enhance the functional robustness of the target systems (or technologies). Orthogonal arrays and experimental-design techniques are very commonly used to plan experiments because of their effectiveness and user-friendliness. After determining optimal settings for control factors, engineers need to compare the robustness of both the optimal and initial settings of the target systems. Finally, confirmation tests need to be conducted to check whether the robustness improvement can actually be reproduced.

The confirmation test is an important stage in robust engineering. After conducting experiments and analyzing the results, engineers are rarely certain of the reproducibility of the results at the optimal setting as compared to the initial setting. Therefore, they should conduct confirmation tests to validate the improved robustness that was achieved at the optimal setting. If the results validate the predicted improvement, the engineers (and managers) can be sure of the effectiveness of robust engineering, and these engineers will almost certainly apply robust engineering in their daily routine later. However, if their first projects fail, they are unlikely to ever use robust engineering methods again. Therefore, it is critical to provide engineers with as much consultation as possible when they conduct their first robust engineering projects; this is the time when robust engineering instructors need to help the engineers select the basic functions and associated robustness measurements of the target system. Many people may believe that robust engineering methods are equivalent to orthogonal arrays and analysis of variance methods, but this is not the case. In fact, orthogonal arrays and analysis of variance methods are not really the major issues in robust engineering. Instead, we need to help engineers focus their resources on how to select and measure the basic functions of the target technology, not on orthogonal arrays.

After popularizing robust engineering methods in the company for several years, the author surprisingly found that the major roadblock to company-wide implementation was middle-level management. Many engineers complained that they were not able to conduct as many robust engineering activities as necessary because their managers did not understand the value or necessity of robust engineering.

In the opinion of the author, some in middle-level management are not really long-term oriented. These individuals focus their resources and attention on fire-fighting activities that solve existing problems instead of focusing on preventing future problems. All in middle-level management should have appropriate training in robust engineering to ensure that they understand the meaning and necessity of robust engineering. In fact, middle-level managers have a responsibility to popularize robust engineering methods in their departments so as to enhance company productivity. When training middle-level managers, it is not necessary to supply sophisticated technical details of robust engineering. Instead, instructors need to focus on illustrating the reasons and necessity for conducting robust engineering, including its role in preventing potential quality problems. The objective of robust engineering training for middle-level management is to convince these managers that preventing problems is much more productive than solving them after they arise. A manufacturing company cannot develop robust technologies unless all levels of management understand the importance of problem prevention and are committed to investing engineering resources to conduct robust engineering activities. If those middle-level managers who are convinced of the value and necessity of robust engineering are later promoted to top-level management positions, the company will have enhanced opportunities to implement robust engineering methods.

Most robust engineering projects are conducted by entry-level engineers. However, understanding, support, and encouragement from management are critical for company-wide popularization of robust engineering. It is not efficient to use a bottom-up approach to popularize robust engineering methods; it is more efficient to have a top-down approach. Robust engineering can be implemented efficiently if top management supports it. On the other hand, if top management is not convinced of the power and value of robust engineering, middle-level management will not be very willing to send their employees to attend the necessary training or workshops on robust engineering, and consequently not many successful case studies will result. Eventually, robust engineering activities will diminish in this case.

## 6.5    IMPACT OF ROBUST ENGINEERING ON THE COMPANY

At the introductory stage, many engineers at Nissan volunteered for training seminars in robust engineering to enhance their knowledge. However, robust engineering did not become a part of the daily routines of these engineers immediately. The major quality-engineering resources of the company were still focused on downstream fire-fighting activities, such as reducing warranty cost or solving existing quality problems that caused customer dissatisfaction. Top management still focused on the most visible quality problems that were reaching the market. This approach is short-term oriented and cannot fundamentally improve the competitiveness of the company. The author discovered the following obstacles to the implementation of robust engineering:

- After initial training, some managers and engineers still did not fully understand the objective of robust engineering, which is to fundamentally improve the basic functions of generic technologies so as to prevent the occurrence of numerous downstream quality problems. These individuals did not take proactive activities to improve the basic functions of the company's products. Their only goal was to reduce customer dissatisfaction or warranty cost.
- Managers and engineers had little experience or knowledge in the implementation of robust engineering. Because they were not capable of applying robust engineering methods, they preferred to stick with traditional quality-engineering methods (e.g., life-cycle tests or validation tests, as discussed in Case Study 2).
- Some engineers did their best to conduct robust engineering projects, but unfortunately received little encouragement or support from management. Thus, they became less proactive and their working attitude changed from new-technology oriented to existing-problem oriented.

Initially 30 robust engineering projects were conducted under the direction of Dr. Taguchi. In 1990, after the robust engineering department was set up, some of these robust engineering projects already had impressive results. Most were conducted by entry-level engineers, who are more willing to adapt to robust engineering methods than senior engineers or management. The results of these projects showed the introduction of robust engineering to be a success. About 200 engineers of the approximately 2000 research, development, and design engineers in the company were involved in these 30 projects. Thus about 90% of engineers and managers were still not involved in the implementation of robust engineering. The 10-year goal of the robust engineering department is to facilitate at least 300 robust engineering projects each year so that all research, development, and design engineers will be involved in robust engineering activities.

In 1991, some engineers began to apply robust engineering to their daily assignments. Unfortunately, their attempts were frequently interrupted by their supervisors or managers, most of whom were still trying to use their engineering resources to solve existing quality problems and who still treated robust engineering as a non-value-added activity. To remove this obstacle, the author set up an introductory robust engineering class for management, to show them the value and effectiveness of robust engineering. Currently, any newly promoted supervisors or managers of the company are required to attend a one-day introductory robust engineering class.

Because of the positive results of the first 30 robust engineering projects, more and more engineers and mid-level managers came to understand the value and effectiveness of robust engineering. In addition, one of the first to volunteer to set up the robust engineering department has been promoted to the board of directors of Nissan Motor Co., Ltd. Robust engineering has been getting more and more management support.

As mentioned above, many engineers and managers did not notice the fast advancement of new technologies and still stuck to technologies that were 5 or 10 years behind the competition's. As a result, the company failed to develop many competitive technologies. The construction of new-technology bases is critical for company survival in this highly competitive market. The author believes that the most efficient way to restructure technology bases is to bring robust engineering far upstream in the total-product-development process so as to develop robust technologies. Compared to a fire-fighting approach, it may take more time and engineering resources to develop robust technologies at such an upstream stage, but the effort will pay off by preventing downstream quality problems and by reducing the product-development cycle.

After several years of the author's promotion of robust engineering, some engineers and managers came to understand its value and effectiveness. More and more managers are willing to send their employees to training or case-study workshops. Currently, robust engineering has the support of top management and has been integrated into the culture of the company. In the experience of the author, it is critical to convince top management of the value and effectiveness of robust engineering. If top management supports robust engineering, middle-level management will encourage their

employees to conduct robust engineering activities. Eventually, robust technologies will be developed gradually inside the company and technology
bases will be constructed.

Currently, more than 100 robust engineering case studies are being
conducted each year in the company, the results and reports of which are
stored in the engineering knowledge center so that engineers can easily refer
to them. These real-life case studies are excellent materials for on-the-job
training in robust engineering.

The following chart (Fig. 6.1) summarizes the number of robust engineering projects conducted at Nissan Motor Co.

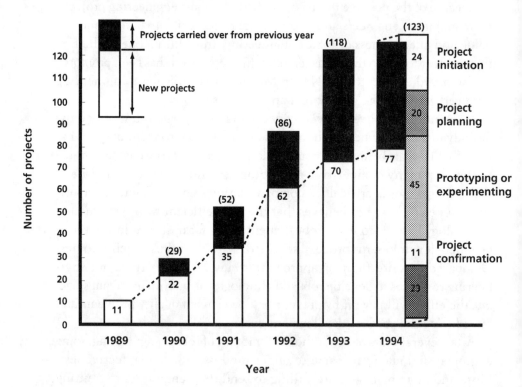

**Figure 6.1** Number of robust engineering projects conducted in the company. The
numbers in parentheses give the total number of robust engineering projects for
each year.

# Classifications of Quality Characteristics and Associated Quality Requirements

## 7.1 DEFINITION OF QUALITY

Robust engineering should be applied primarily in the research, development, and design departments of a manufacturing company. The objective of robust engineering is to solve quality problems fundamentally and efficiently through design activities. However, the term "quality" needs to be clearly defined by a company, since engineers and managers do not necessarily have a common definition of the word. The definition of "quality" may differ from company to company and industry to industry. It is a challenge to create a common definition of quality for all engineers and management. Some definitions are based on the value of products or services, others on subjective judgments or customer satisfaction. In robust engineering, the "quality loss" of a product is an estimate of the total monetary loss imparted by the product to society from the time the product is shipped from the manufacturing plant.

The technical definition of quality may not be easy for engineers and management to accept immediately because every person has his/her own definition of quality. However, to enhance the productivity of the company, we need to have a common definition of quality. A common definition is critical for product-planning and marketing activities, which in turn are vital to the survival of the company. Product-planning and marketing activities should focus on making products meet customer requirements. These activities are related to the evaluation of downstream customer quality.

However, improving the quality of a target product is very different from evaluating its downstream quality. To enable the engineers and managers of the company to understand the definition of quality, the total quality loss of a product can be further divided into the following two terms:

- Functional loss: the monetary loss due to functional variation of the product;
- Usage loss: the monetary loss due to negative side effects, including initial purchase and usage cost.

The definition of quality in terms of quality loss is based on the monetary loss of a product throughout its operational life. This definition is very different from the traditional definition of quality, which is essentially based on the very subjective judgments of customers. The acceptance of the definition of quality loss throughout the company is vital for company-wide implementation of robust engineering.

## 7.2    DOWNSTREAM QUALITY AND UPSTREAM QUALITY

To help participants in the robust engineering workshops understand this definition of quality, the instructors asked them the following question:

> What is the quality of the service provided by the Tokaido Bullet Train (an express train running between Tokyo and Shin-Osaka)?

After asking this question, the instructors did a quick survey of the attendees. There were usually 50 people in each workshop. In general, about 70% (35 people) considered the Bullet Train to be of high quality, 20% (10 people) considered it to be of low quality, and 10% (5 people) thought its quality to be neither high nor low. Next, the instructors asked the attendees how they judged the quality of the Bullet Train. According to the survey, about 70% of the judgments were based on participants' personal experience with the Bullet Train. The reasons for high-quality or low-quality opinions of the Bullet Train were as follows:

The quality of the service provided by the Bullet Train was considered to be high because of:

- High speed
- Generally on-time arrivals
- Few accidents
- Comfortable ride
- Quietness
- Provides affordable and efficient transportation for the public

The quality of the service provided by the Bullet Train was considered to be low because of:

- Frequent delays caused by snow or rain
- Occasional late arrivals
- Heavy vibration
- Not being as convenient as personal vehicles

From this survey, we can see that while some people have very positive opinions, others have negative opinions about the Bullet Train. If we study these in more detail, we find that they are, in fact, all visible or palpable downstream symptoms caused by variations in the basic functions of the Bullet Train. From the standpoint of robust engineering, the most important issue is to improve the robustness of the basic function of the Bullet Train, not to eliminate the associated downstream symptoms. In other words, we need to determine what the basic function of the Bullet Train is and then apply robust engineering to make it robust against various, extreme environmental conditions.

The objective function of the Bullet Train is to transport great numbers of people from one city to another in a speedy and reliable manner. In this definition, the target values for the number of people and the speed of the train may vary with societal and environmental changes from time to time. If the Bullet Train performs its objective function, it will always be on time. In order to achieve this objective function, the Bullet Train needs to perform its basic function (i.e., translating electric power supply into train velocity) consistently and reliably. Unfortunately, few engineers in the robust engineering workshops ever think of the Bullet Train example from the viewpoint of its basic function. From this example, it can be seen that most engineers focus their daily activities on solving downstream quality problems. In other words, few ever think about improving the robustness of the basic functions of their products or systems. Most engineering activities still focus on eliminating downstream quality problems of current products or systems, one major obstacle to the advancement of quality engineering in the company. To restructure the technological basis of the company, we need to shift the paradigms of engineers and managers from solving downstream quality problems to enhancing the robustness of the basic functions of their new products or systems.

As for the Bullet Train, in reality, it was frequently delayed because of snow or rainy weather. More than 500,000 passengers were delayed on a single day because of one big storm. If we assume that the average income of these passengers is 50 million yen per year and that each passenger works 250 days per year, this one-day delay of the Bullet Train would impart a monetary loss of 10 billion yen (less than $100,000,000) to society. This monetary loss is caused by the instability (i.e., lack of robustness) of the basic function of the Bullet Train.

## 7.3   QUALITY AND FUNCTIONALITY

The momentary loss caused by the variation (i.e., instability) of the basic function of the target system is called functional variation loss. However, quality is a very different issue from functionality. For example, the Bullet Train may be able to achieve its function by transporting great numbers of people from one city to another city in a speedy and reliable manner. However, it may also cause a lot of disturbing vibration and noise for local residents. If this is so, then we may conclude that the functionality of the Bullet Train is excellent, but its downstream quality is very low (because it causes a lot of downstream side effects.) Indeed, some engineers in the robust engineering workshops commented that the basic function of the Bullet Train is excellent, but the quality provided by the service of the Bullet Train is low.

The objective of traditional quality engineering is to eliminate all negative side effects so as to satisfy customers. However, from the viewpoint of robust engineering, the objective of quality engineering is to fundamentally enhance the robustness of the basic function of the target system so as to prevent the occurrence of negative side effects. In order to do this, the first step is to develop a reliable measurement technique for the system's functional robustness. It is inefficient and time consuming to measure all negative side effects and then try and eliminate them one by one.

To sum up, the quality of a product is estimated as the total loss imparted by the product to society from the time the product is shipped from the manufacturing plant. However, this quality loss is caused primarily by functional instability (i.e., variation) in the basic function of the product. Eliminating downstream quality problems is not an efficient way to improve a product's quality. The most efficient way to improve the quality of a product is to enhance its functional robustness at a very early development stage.

## 7.4   CLASSIFICATION OF QUALITY CHARACTERISTICS

In robust engineering, quality characteristics can be classified into the following four categories:

1. **Origin quality characteristics:** These quality characteristics are related to the functional robustness (i.e., repeatability, stability) of product or

process technologies. Examples of origin quality characteristics are dynamic S/N ratios. Origin quality characteristics are frequently applied to evaluate the functional robustness of target product or process technologies. Three factors affect these quality characteristics: technological readiness, flexibility, and repeatability.

2. **Upstream quality characteristics:** These characteristics are related to the deterioration of key requirement characteristics from their target specifications. Upstream quality characteristics are applied to optimize the design of specific products or processes. Static S/N ratios are examples of these quality characteristics.

3. **Midstream quality characteristics:** These quality characteristics apply to the manufacture or assembly of specific products. Dimensional nominal values and tolerance specifications of design drawings are examples of midstream quality characteristics. Midstream quality characteristics are frequently used as quality-control characteristics in manufacturing/assembling plants.

4. **Downstream quality characteristics:** These are also called customer quality characteristics. Examples of downstream quality characteristics are failure, vibration, audible noises, fuel economy, pollution, etc. Administrators and managers at manufacturing companies often use such quality characteristics to measure whether their products meet the requirements of customers.

Of the four types of quality characteristics above, customers are sensitive only to the influences of downstream quality characteristics, as illustrated in the Bullet Train example. Customer quality characteristics are, in fact, symptoms of the deterioration of associated upstream characteristics. Administrators and managers of manufacturing companies tend to collect downstream quality data directly from their customers to determine whether or not their products meet customer needs or requirements. Of course, administrators and managers expect quality problems in their products to be as few as possible. However, development engineers should not apply customer quality characteristics to evaluate the quality of their new technologies or products, as customer quality characteristics are only the symptoms of the deterioration of associated upstream quality characteristics. It is inefficient to use downstream quality characteristics to develop new technologies or products because doing so diverts development engineers from applying upstream quality characteristics to fundamentally solve quality problems. What engineers can do is eliminate all the negative downstream quality characteristics they discover when using numerous exhaustive validation tests.

Development engineers need to focus their engineering resources on enhancing the robustness of the basic functions of their new technologies or products instead of eliminating associated downstream quality problems. If the basic function of a new technological system or product is very robust, the energy-transformation efficiency of the system or product will be very high. As a result, most input energy will be transferred to perform the intended basic function, while little input energy will be transferred to cause negative side effects or downstream quality problems. Eliminating downstream quality problems is not an efficient way to fundamentally solve quality problems. The most fundamental and efficient way to solve quality problems is to improve the functional robustness of a technology or specific product at the early stages of the product-development process.

## 7.5 CUSTOMER REQUIREMENTS AND QUALITY CHARACTERISTICS

In the very early stages of a technology- or product-development process, engineers should focus their resources on upstream quality characteristics, not on downstream quality characteristics. By focusing on upstream quality characteristics, engineers will have more opportunity to design quality into a new technology or product, resulting in more efficient quality engineering and reduced development-cycle time. This is the reason why development engineers need to bring quality engineering to the most upstream stages of the product-development process. The following three factors are critical to the development of robust technologies:

1. **Technological readiness:** Before the target values of downstream customer quality characteristics are specified, development engineers need to conduct robust engineering activities to enhance the functional robustness of their new technologies. In other words, technology-development activities need to be conducted *before* product-planning or product-marketing activities.

2. **Flexibility:** One generic technology should be applicable to a family

of new products so as to reduce average development cost. In addition, the generic technology should be applied to developing a wide range of products and thereby meeting a wide range of downstream customer requirements.

3. **Repeatability (robustness):** At a very early stage in the product-development process, the functional robustness of the target generic technology must be enhanced so as to ensure the stability and reliability of the associated downstream products.

To develop robust technologies, engineers need to focus their attention and resources on the basic functions of their target systems. In addition, they need to consider how to measure this functional robustness. Continuous development of robust technologies is the most efficient way to restructure the technological bases of a manufacturing company.

## 7.6   THE RELATIONSHIP BETWEEN QUALITY AND COST

The relationship between the quality and cost of a product should be determined before its development. The goal of any manufacturing company is to make a profit. The total cost of a product contains both the manufacturing cost and the maintenance/usage cost. In order to make a profit, a manufacturing company needs to reduce the total cost as much as possible. One common way to reduce the manufacturing cost is to loosen the tolerance specifications of design parameters or to use lower-grade materials/components. However, this method negatively impacts the quality of the associated products. In other words, if development engineers improve the quality of a specific product by tightening the tolerance specifications of design parameters or by using high-quality components/materials, the manufacturing cost of the product will increase accordingly.

On the other hand, the goal of robust engineering is to improve the quality of a product while reducing the manufacturing cost. If robust engineering activities are competently conducted, high-quality products can be manufactured using low-grade components or materials. In other words, a robust product should be able to tolerate the variations in low-grade materials or components. High-quality products can thereby be manufactured at low cost. A robust product should be reliable and durable under various usage conditions, so the maintenance cost of the product can be reduced too. Eventually, we can minimize the total cost of the product without sacrificing its quality.

The purpose of confirmation tests for a new product is to judge whether or not potential customers will be satisfied with it. However, these judgments are usually very subjective. If test conditions differ from customer-usage conditions, these confirmation tests will be misleading and may produce incorrect conclusions. At the early development stages of a new product (or new technology), we need to focus our resources on the robustness of the product (or technology) instead of its downstream quality characteristics. Confirmation tests can be conducted at downstream development stages to check whether or not potential customers will be satisfied.

## 7.7  PREVENTING PRODUCT FAILURES BY USING ROBUST ENGINEERING

The marketing survey for a new product can be conducted only after the product is shipped to market. However, at this stage, we may not have enough time to conduct an accelerated life test to estimate the reliability and durability of the product under all possible customer-usage conditions. Even worse, we do not have sufficient design freedom to improve or correct the design because all design parameters are already specified. An alternative and more efficient way to improve the quality of a future product is to conduct robust engineering at the very early stages of the product-development process.

At the very early stages of the product-development process, development engineers usually have a very limited number of prototypes on which to conduct validation tests. Fortunately, at these early development stages, engineers have much more freedom to design robustness into products or new technologies (i.e., to reduce functional variation in the basic function of the target systems) than they do at downstream stages. In other words, the further upstream robust engineering is implemented, the more opportunities development engineers have to enhance the functional robustness of target products or technologies.

The goal of traditional reliability-engineering methods is to determine the failure modes of products and then to predict their operational lives as accurately as possible. In fact, this aim is somewhat beside the point, because predicting the lives of products does not really improve their reliability or quality. It is wasteful to invest quality-engineering resources to conduct life-cycle tests. Traditional reliability-engineering activities are essentially counting activities and do not really improve the basic functions of the target products. Improving the functional robustness of the target product or technology at a very early stage of the product-development process is more efficient than conducting numerous downstream reliability-engineering activities or validation tests. Validation tests are only useful for checking whether or not the manufactured products meet customer requirements and needs. However, at the validation stage of the product-development process, most design factors or variables are already specified and little design freedom is left for the improvement of functional robustness. In the opinion of the author, validation tests are merely counting activities, not constructive development activities. Only at the very early stages of the product-development process do development engineers have enough design freedom to enhance the functional robustness of new products or technologies.

## CASE STUDY 2: WELDING STRENGTH AND LIFE CYCLES

In traditional reliability engineering, it is common to apply life-cycle tests to estimate the operational lives of specific products. In this case study, operational life is estimated by applying robust engineering methods to measure the deterioration of objective quality characteristics, rather than the life cycles, of laser-welded parts. This case study was conducted by the Advanced Machining Technology & Development Association of Japan and was presented in the Symposium of Quality Engineering for New Materials and Associated Machining Processes, 1991. The robust engineering methods illustrated in this case study are much more meaningful and time efficient than traditional life-cycle tests.

In a typical life-cycle test, engineers apply a load and conduct repeated operations on a part until the part breaks down or fails. The number of the operational cycles is used to estimate the life of the part. Generally speaking, the life of a typical automobile part is supposed to exceed 10 million cycles. If we reduce the load on the part, the life cycles may increase several times. Obviously, it takes significant testing time and resources to determine the life cycles of a given part. (Assuming that we can conduct 10 cycle tests per second, it will take 280 hours to conduct 10 million tests. If we can test continuously day and night, it will take 12 days to conduct the test.)

The purpose of a life-cycle test is to check whether or not a specific part meets the life-cycle specification. Life-cycle tests are usually so time and resource consuming that testing engineers can conduct only a very limited number of them on prototypes of a specific design. As a result, test results seldom represent the variation in the life cycles of all mass-produced parts of the same design. For example, say we conduct life-cycle tests on two proto- types of design A and also on two prototypes of design B. After the tests are done, we compare the data of designs A and B with the life-cycle specification to see whether they correspond. As shown in Fig. 7.1, the life cycles of the two prototypes of design A are above specifications; thus, design A is judged to be "acceptable." The life cycles of the two prototypes of design B are below specifications; thus, design B is judged to be "unacceptable." However, these test results may not hold true for mass-produced versions of designs A and B.

Because, at the very early stages of the new-product development process, engineers have only a very limited number of prototypes, they are not able to determine the mean value and variation of the life-cycle distribution of mass- produced versions of a new design. If the objective quality characteristic of mass- produced versions of a design is of little variation, validation test data may be useful. Unfortunately, most mass-produced products have significant variation, and therefore, the life-cycle data of a design based on tests of several prototypes seldom represent the real operational lives of the mass-produced versions of the same design. Assuming that the distribution of the life cycles of mass-produced products of design A and B are as illustrated in Fig. 7.1, we can see that the results of the validation tests of the prototypes are somewhat misleading.

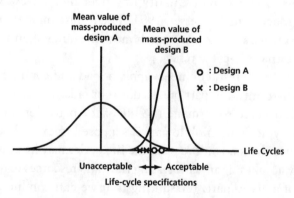

**Figure 7.1** Life-cycle distributions for designs A and B.

Assume that the quality of the two prototypes of design A is much higher than that of mass-produced products of the same design, as shown in Fig. 7.1. The life cycles of the two prototypes of design A exceed specifications; consequently, all products of design A will be judged to exceed specifications by test engineers. From Fig. 7.1, we know that this judgment is inappropriate. If we put design A into mass production and deliver the products to the market, many downstream quality problems will arise, since most mass-produced products of design A do not meet the specifications. However, because only a very limited number of prototypes were tested and these exceeded the specifications, test engineers may come to an incorrect conclusion. Conversely, assume that the quality of the two prototypes of design B is much lower than the average mass-produced products of the same design. According to the curve shown, most mass-produced products of design B exceed life-cycle specifications, and therefore are acceptable for market conditions. However, because of the two inferior prototypes of design B, all products of design B are judged to be unacceptable. As a result, excessive engineering resources may be wasted on improving design B. In other words, products of design B may be overengineered because of misleading test results.

The major point of a validation test is to estimate the life cycles of mass-produced products as accurately as possible. However, it is difficult to do so when using a very limited number of prototypes. Some engineers may apply the life-cycle data of current products to estimate the life-cycle distribution for mass-produced products of a new design. This approach is also inappropriate because the life-cycle distribution of the new design may be very different from that of the old design if the new design, although similar in appearance, involves different components or manufacturing processes.

Another problem with life-cycle tests is that their only testing characteristic is the number of life-cycles of the tested prototypes. Other quality characteristics such as surface finish, wear, structural integrity, audible noise, etc., are usually neglected. In fact, users of products (e.g., car brakes) often apply quality characteristics (e.g., noise, vibration, braking distance) other than life-cycles to judge whether the products are still usable or not. Most drivers will have their brakes repaired long before they fail completely. Currently, many life-cycle tests are conducted by robots or other automatic machines. Consequently, only the number of life-cycles are recorded while other quality characteristics (e.g., wear, deterioration) are often ignored. In fact, using the results of traditional validation tests, testing engineers can seldom determine the reasonable operational lives of mass-produced products.

If the life-cycle variation of mass-produced products of a certain design is large, development engineers need to strengthen the design to ensure that all mass-produced products of that design meet the life-cycle specification. Consequently, the weight and the manufacturing cost of the design will both increase. A more efficient way to ensure that all mass-produced products of a design meet the specification is the two-step, robust engineering procedure, which involves first reducing the variation of objective quality characteristics and then adjusting the mean value of the characteristics to meet the specifications.

In this case study, the author illustrates how to apply robust engineering methods to estimate the operational life of prototypes in a more time-efficient way than traditional life-cycle tests. Simply put, in this case study, testing engineers used the deterioration of the elasticity of welded parts, instead of the number of life cycles, as the measurement characteristic. Using a microscope to study the cross section of fatigued parts (or products), we usually find that most failure/fatigue modes are caused by physical or chemical deteriorations of key design parameters. If these physical or chemical deteriorations exceed certain limits, the parts (or products) will not be able to perform their intended function. In other words, the objective quality characteristics of most parts (or products) will deteriorate if they are used repeatedly. For example, say we extend and release a rubber band repeatedly. At the early cycling-test stage, the rubber band may have a consistent elasticity. However, after a certain number of operational cycles, its elasticity will deteriorate significantly. Eventually, the rubber band becomes fatigued and unable to perform its basic function (i.e., provide elasticity). We can measure the deterioration of the elasticity, instead of total life cycles, of the rubber band to estimate its operational life.

This method can be applied similarly to those parts (or products) that are under repeated static, bending, or dynamic loads. In short, in life-cycle tests we want to measure the deterioration of the elasticity (Young's modulus) of the objective parts (or products). Elasticity is defined as the ratio between stress (load) and strain (or yield). Ideally, the elasticity of a part (product) should be constant throughout its operational life. However, because of the material fatigue described above, this ratio gradually deteriorates. The deterioration of Young's modulus will be the measurement for the functional robustness of the part under repeated operational conditions.

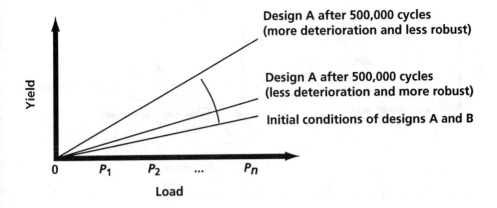

**Figure 7.2** Deterioration of elasticity (load-yield functionality).

According to Hooke's law, the relationship (i.e., function) between the load applied to a tested part and the yield of the part should be linear (Fig. 7.2). We want this linear functionality between load and yield of a test part to vary little throughout the part's operational life. This load-yield functionality is, in fact, the elasticity (i.e., spring constant) of the part. If the elasticity of a part shows little variation under repeated operations, it will have a very long operational life. Conversely, if the elasticity of a test part shows large variations, it is not robust and will not be able to perform its basic function consistently. Therefore, we can use the deterioration of the elasticity of test parts to estimate their life, conducting only a small fraction (around 1/10) of the life-cycles normally used. For example, it may require 10 million cycles to cause a test part to completely fail. However, if we apply the deterioration of elasticity as the measurement characteristic, we probably only need to conduct 100,000 to 1 million cycles to get enough data to estimate operational life (Fig. 7.3). More important, this deterioration data of specific part/material can be used as a knowledge base for the development of future product technologies. If we have such information at the technology-development stage, we need not go through exhaustive life-cycle tests for each material/part in new-product-technology development. As a result, the development time for new products or technologies will be reduced significantly. After a specific product is developed from newly developed technologies, we can conduct a confirmation life-cycle test for the new product.

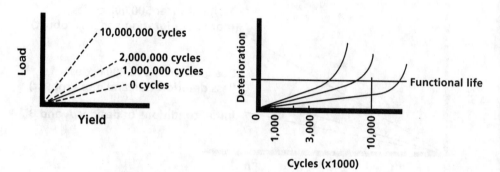

**Figure 7.3** Deterioration of the elasticity in a confirmation test.

We applied the methods described above to evaluate the welding strength of specific parts processed by a laser-welding machine and then to develop robust laser-welding technology. We began by first discussing the objective of this project with the engineers responsible for developing laser-welding technology. The objective was to determine optimal welding conditions so as to maximize the operational life of the welded parts. However, the engineers had no information about the variation of the welding strength of welded parts. In fact, it would take a great deal of time to determine such variation. We used 0.8-mm steel plates and 4-mm plates to conduct this robust engineering project. Plates of 0.8-mm are used primarily for vehicle body panels. The two most commonly used welding methods for 0.8 mm plates are butt weld and fillet weld. In comparison, the most commonly used welding method for 4-mm-thick plates is the T-type weld. Figure 7.4 shows the three welding types and the loads on test parts.

First, we drew up load-yield charts, as in Fig. 7.6. According to Hooke's law, the functionality between load and yield should be linear. The deviation of the elasticity of a test part from its linear proportionality is defined as its deterioration percentage. Testing engineers applied certain loads to the test parts and measured their deterioration percentage at 10,000, 20,000, 50,000, 100,000, 200,000, 500,000, and 1,000,000 cycles. One 4-mm test plate broke at 20,000 cycles under a heavy load. Test engineers therefore reduced the load and conducted the life-cycle test again.

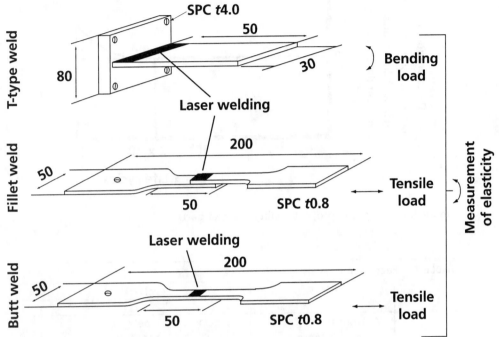

**Figure 7.4** Three welding methods and loads on test parts.

Next, testing engineers collected the test data and charted it, as in Fig. 7.5, where deterioration in the proportionality constants of test parts are plotted. Generally speaking, after numerous (e.g., 200,000 in Fig. 7.5) cycles, the proportionality constants of most parts will increase (i.e., deteriorate) significantly. In other words, the elasticity functionality of test parts will deteriorate significantly after numerous operational cycles. From these charts, we can see how proportionality constants (between load and yield) of test parts vary with the number of life cycles. In most life-cycle tests, we tested the parts up to 500,000 cycles.

In this project, we selected 8 control factors to enhance the robustness of the laser-welding technology. We also had two noise factors. These control and noise factors are listed in Table 7.1. The control factors were assigned to an $L_{18}$ orthogonal array; thus, we had 18 combinations for control-factor levels. Testing engineers ordered test parts for these 18 combinations, then conducted life-cycle tests and measured the proportionality constants (between load and yield) of these test parts, as shown in Fig. 7.6.

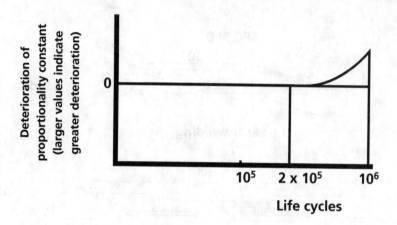

**Figure 7.5** Changes in proportionality of a test part.

**TABLE 7.1** Factors and levels

| | Factors | Levels | | |
|---|---|---|---|---|
| | | 1 | 2 | 3 |
| | A: Laser type | Type 1 | Type 2 | |
| | B: Lens distance | 190.5 | 254 | 190.5 |
| Control | C: Laser-power output | -10% | Standard | +10% |
| factors | D: Welding speed | -10% | Standard | +10% |
| | E: Gas variety | Ar | He | $N_2$ |
| | F: Gas-flow rate | 30 | 40 | 50 |
| | G: Gas-flow direction | Center & boundary | Center | Boundary |
| | H: Lens position | -3 | -1.5 | 0 |
| Noise | K: Test part shapes | $K_1$ | $K_2$ | $K_3$ |
| factors | N: Test cycles | 0 | 500,000 | |

At the beginning (0 cycles) of the life-cycle tests, the proportionality has little variation between $N_1$ and $N_2$, while after 500,000 test cycles, this proportionality constant varies significantly. The variation of the proportionality constant between $N_1$ and $N_2$, ranging from 0 to 500,000 cycles, was used to evaluate the functional robustness of the welding condition of each combination of control-factor levels. The welding condition with the minimum amount of functional variation is the most robust welding condition.

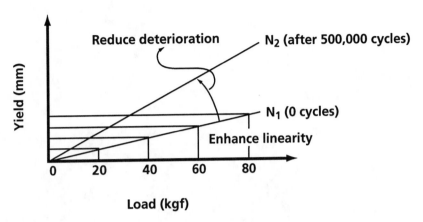

**Figure 7.6** Load-yield functionality.

During these robust engineering experiments, 4 test parts out of the 18 combinations of the $L_{18}$ array broke before 500,000 cycles. There are two possible causes for this: one is overload on the test parts and the other is inferior welding condition. Test engineers examined these broken test parts and concluded that their breakage was due to inferior welding conditions. Thus, though some data were missing because of these broken parts, we could nevertheless conclude that the welding conditions of these parts were not optimal. The missing data were estimated using iterative numerical methods.

After the experiment, the optimal welding setting was estimated to exceed the initial welding setting by 17 decibels, which meant that parts processed using the optimal setting should last 7 times longer. Test engineers conducted confirmation tests to compare the optimal and initial welding settings. These tests confirmed the estimations.

In summary, using robust engineering methods to conduct life-cycle tests, we can quantify the deterioration of the strength of laser-welded parts. We can also apply a similar approach in the development of new metal materials, the welding strength of which can be measured using nondestructive methods. This case study is a good example of the application of robust engineering in the development of welding and material technologies.

# The Basic Function of a Technology

## 8.1 THE FUNDAMENTALS OF QUALITY PROBLEMS

To fundamentally solve a quality problem, we need to focus engineering resources on the problem's most upstream causes so as to desensitize the target system against them. The alternative is to eliminate the symptoms downstream, but this is not an efficient way to solve a quality problem.

To identify the most upstream causes of a quality problem, we first need to consider the basic function of the target technology (or specific product). The basic function of a technology is what the technology or product is supposed to do to fulfill its purpose. The focus of robust engineering is on how to improve the robustness of the basic function of the target technology, not the associated downstream quality symptoms. To practice robust engineering, engineers need to shift their paradigms from eliminating downstream quality problems to enhancing the robustness of the basic function of a generic technology or specific product.

In robust engineering applications, one big challenge is to determine the basic functions of the target technology (or product). The challenge exists because the basic functions of some engineering applications may not be obvious. In fact, design engineers may know the basic functions of the target technology (or product) best because a precise and complete understanding of the technology's purpose and objectives is needed to convert design concepts into design drawings. However, most other engineers have not had this training and may have some difficulty in understanding the basic functions of technologies (or specific products). It is therefore a good idea to have design engineers explain the basic functions of the target technology (or product) to other team members at the initial stages of robust engineering projects.

To determine the basic function of a given technology or a specific product, we must identify its objective and decide which engineering methods will best achieve it. The most efficient way to train engineers to determine basic functions is through real-life case studies. Below are several real-life examples that illustrate how to enhance the robustness of the basic functions of the target technologies.

## 8.2 BALL JOINTS AND FRICTION FORCE

The structure of a ball joint is illustrated in Fig. 8.1. It is composed of a steel ballstud and an aluminum socket with a spherical cavity. The fringe of the socket is bent inward to hold the spherical head of the ballstud in place. The basic function of a ball joint is to translate force from one link to the other. An ideal ball joint should have no friction and its two links should be able to rotate freely.

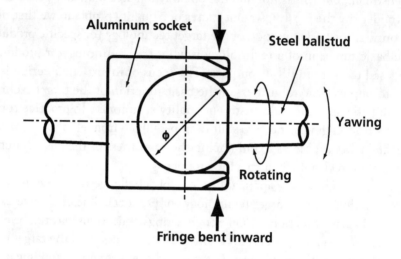

**Figure 8.1** Structure of a ball joint.

The ball joints from a certain supplier often showed excessive friction. At the time, process engineers in ball joint engineering were working on the fringe-bending process to reduce the average friction force. They evidently considered the fringe-bending process to be the deciding factor in ball joint friction. However, they did not have much success in reducing the friction force, so they asked the author to help them apply robust engineering to optimize the manufacturing processes. Discussion with the engineers revealed that they had not spent much time or resources in studying the machining processes associated with ball-joint ballstuds or sockets.

As mentioned above, the friction force of a ball joint should be as little as possible. If too great, the ball joint may not be able to transmit force precisely from one link to the other, causing downstream quality problems. In other words, the ball joint may not be able to predictably perform its basic function. The supplier's engineers usually measured friction force by yawing or rotating one link of the ball joint against the other. In many cases, the friction force was so great that they could not even rotate the links by hand. If installed in automobiles, these defective ball joints would cause significant downstream quality problems. Therefore, friction force is a downstream quality characteristic of ball joints, one that needs to be minimal.

At that time, quality-by-inspection activities were used to control downstream quality problems of ball joints. This involved a tremendous amount of human resources to yaw and rotate all manufactured ball joints. Those ball joints with acceptable friction force were kept, and the remainder were scrapped. This type of quality control is very time consuming and costly, and the author concluded that the engineers had not yet fundamentally solved the ball joints' quality problems. The author recommended that the engineers conduct experiments to identify the most upstream causes of the excessive friction force.

## 8.3 MACHINING A PERFECTLY SPHERICAL HEAD AND CAVITY

The author wondered what was actually causing the friction force, and to solve the problem fundamentally, kept asking the engineers about the most upstream cause. In fact, the engineers had never thought about where the friction force of ball joints actually occurred. However, they did realize that ball joints of all sizes showed significant friction. In ball joint manufacturing processes, the spherical head of a ballstud is inserted into the spherical cavity of an aluminum socket. Next, the fringe of the socket is bent inward to hold the spherical head of the ballstud, as shown in Fig. 8.1. The socket material is aluminum and is supposed to spring back slightly after the bending force is released. Thus, it seemed to the author that the bent fringe should not impart any friction force to the spherical head of the associated ballstud, and therefore, the socket-fringe bending process was not likely to be a root cause of the friction.

At first, the engineers focused on the fringe-bending process as a possible cause for the excessive friction. After further discussion, it was concluded that the friction might be due to imperfections in the sphericity of the ballstud heads and socket cavities. In design drawings for ball joints, ballstud heads and socket cavities are specified as being perfectly spherical. However, because of various manufacturing imperfections and deterioration factors, neither head nor socket is ever perfectly spherical. In fact, it is still very difficult to use current measurement techniques to measure the sphericity of ballstud heads and socket cavities. So, although the friction-force problem was unlikely to be closely related to the fringe-bending process for aluminum sockets, it conceivably might stem from the imperfect shapes of ballstud heads and their associated socket cavities. The engineers agreed that such manufacturing imperfections would cause the friction seen between ballstud heads and socket cavities and were ready to take action to control ballstud head and socket cavity sphericity .

If the engineers had continued working on the fringe-bending process, they might never have solved this problem. In other words, if they had not determined the most upstream cause of this friction-force problem, they would never have been able to solve it. The problem was not related to design activities, because design engineers had specified the socket cavities and ballstud heads to be perfectly spherical. Instead, the problem was likely due to manufacturing imperfections in socket cavities and ballstud heads. The most efficient way to solve such a problem is to improve the machining processes of both socket cavities and ballstud heads by focusing on machining them as spherically as possible. The basic function of a machining process is to accurately and consistently transform dimensional specifications into actual product dimensions. If the machining process performs its basic function predictably, the sphericity of socket cavities and ballstud heads will be almost perfect.

As engineers, we need to focus on improving the basic function of the machining processes, not on reducing such downstream quality characteristics as friction force. Improving downstream quality characteristics cannot fundamentally solve quality problems.

The author and the engineers set up a three-year plan to enhance the basic function (i.e., transformability from input signals, or nominal values for product dimensions, into actual product dimensions) of the ball-joint machining processes. The problem-formulation procedure for this project is described in Case Study 3. Another project related to the transformability of product dimensions is Case Study 4: Development of CFRP (Carbon-Fiber-Reinforced Plastic) Injection-Molding Technology.

## 8.4    SQUEAKING AUTOMOTIVE BRAKE PADS

This section discusses how to solve squeaking in automotive brake pads. When drivers step on the brake pedal of an automobile, they often hear squeaking, caused by high-frequency friction between brake pads and associated brake discs. Many drivers dislike these squeaking noises and complain of this problem frequently. Automobile manufactures have deployed significant engineering resources and tried several approaches to solve this quality problem. One common solution has been to change brake pad materials, using various additives. One automobile manufacturer, Volkswagen Motor Co., has developed more than 20 additives for brake pads. The engineers there made numerous brake pads containing these additives and conducted validation tests to select those that caused less squeaking. Next, they conducted durability tests on the selected brake pads. They also measured braking efficiency to check whether they met the braking specifications. Unfortunately, brake pads of high-braking efficiency usually squeak a lot and seldom last very long. Conversely, brake pads that do not squeak are usually low in braking efficiency and sometimes do not even meet braking specifications. Brake engineers are often forced to trade off brake noises and braking efficiency and durability. Although automobile manufacturers have devoted considerable resources to brake pad design, improvements in quietness, braking efficiency, and brake pad durability are still very limited. Some brake engineers even went so far as to state that the performance of automobile brake pads must have reached its upper limits. However, the author thought that the problem was due to low efficiency in the development of brake pad technology. Brake engineers usually focus their attention on very downstream quality characteristics such as squeaking, durability, etc., and not on the basic function of brake pads; thus, the development efficiency of brake pad technology is very low. In the development of any product or process technology, we need to focus on the improvement of basic functions.

The purpose of a brake system is to stop an automobile. In other words, the basic function of a brake system is to convert the dynamic energy of a moving automobile into thermal energy through friction between brake pads and associated brake discs. The thermal energy is supposed to radiate away. If energy-transformation efficiency is high, braking efficiency will be very high and predictable. Figure 8.2 illustrates the relationship between

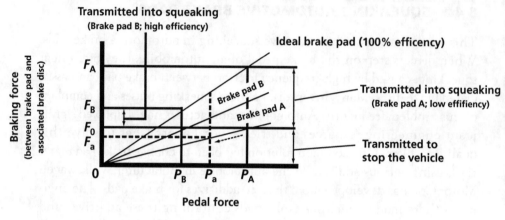

**Figure 8.2** Relationship between pedal force and braking force.

pedal force and braking force. We can treat it as the basic function of the brake system. If the linearity of the basic function is high, the performance of the brake system will be very robust (i.e., predictable).

Assume that the required braking force (between brake pad and associated brake disc) to stop a moving automobile within a certain distance is $F_0$. Also assume that the braking efficiency of brake pad A is low. Thus we need a braking force $F_A$, much higher than $F_0$, to stop the automobile. As is shown in Fig. 8.2, the $F_0$ portion of the braking force $F_A$ is converted into friction energy to stop the automobile, but the remaining force (i.e., $F_A - F_0$) is converted into various side effects, such as squeaking, vibration, excessive wear, etc. To reduce the squeaking in design A, we may reduce braking force to $F_a$ by reducing pedal force from $P_A$ to $P_a$. However, braking force $F_a$ is less than $F_0$ and may not be able to stop the automobile within a predetermined distance. In other words, reducing pedal force may reduce squeaking but may not stop the automobile efficiently. We will have reduced one downstream quality problem but created another. On the other hand, if we increase the braking force of design A by adding more pedal force (or more power assistance), we may be able to stop the car more efficiently, but more useless energy may be converted into negative side effects, and the squeaking problem may even worsen. This may be why high-efficiency brake pads usually squeak loudly.

Assume that the braking efficiency of brake pad B is higher than that of A. Thus, we need only pedal force $P_B$, much lower than pedal force

$P_A$, to generate braking force $F_B$ to stop the automobile within the required distance. As a result, we can reduce the size, weight, and thus the manufacturing cost of the brake system. Because we can minimize the size of the brake system, we can increase the space around it, thereby increasing the thermal-radiation efficiency. As a result, design B will transmit more energy into braking force, and less energy into squeaking, than design A and will therefore be more efficient and quieter than design A.

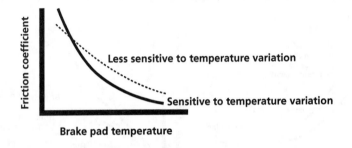

Figure 8.3 Temperature effect on brake pads.

At low temperatures, the friction coefficient is much higher than at high temperatures; thus, the energy-transformation efficiency of brake pads at low temperatures is much higher than at high temperatures. Accordingly, more energy will be transmitted to stop the vehicle and we need not apply much pedal force. As a result, less energy is transmitted into negative side effects such as vibration, squeaking, and so on. On the other hand, at high temperatures, the friction coefficient of brake pads is very low. Thus, we need to apply greater pedal force to stop the vehicle; because the energy-transformation efficiency of brake pads is very low, more energy is transmitted into negative side effects.

As shown in Fig. 8.3, when brake-pad temperature increases, the friction coefficient of brake pads decreases significantly. Brake suppliers had indeed indicated that the friction coefficients of brake pads would decrease significantly when the temperature of brake pads rose above a certain limit. It was believed to be almost impossible to desensitize the friction coefficient of brake pads against temperature. Because brake pads usually have a low friction coefficient at high temperatures, brake engineers must increase brake pad size to ensure that they provide sufficient braking force at high temperatures. However, larger brake pads may generate too much braking force at low temperatures. Thus, brake pad performance is very sensitive to variations in temperature. From the viewpoint of robust engineering, the energy-transformation efficiency of these brake pads shows excessive variation. Eventually, considerable energy will be transmitted into negative side effects, such as vibration, squeaking, etc., especially at low temperatures. Further, the pedal force required to stop the automobile may vary significantly with brake pad temperature. Thus, these brake pads are not reliable and may generate downstream quality problems such as squeaking.

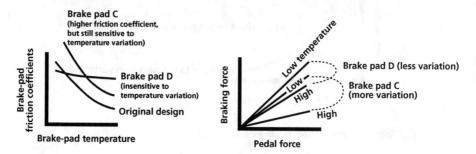

**Figure 8.4** Temperature effect on different brake-pad designs.

As shown in Fig. 8.4, if we increase the friction coefficient of a brake pad without desensitizing it against temperature variation, the braking efficiency of the new design (design C) will be higher than the original design. Brake pad C is correspondingly much more energy efficient and quieter than the original. However, because high-friction brake-pad materials are usually very soft, the pads may not last very long. This is why high-efficiency (i.e., high-friction coefficient) brake pads such as design C usually have poor durability.

From the above, we can conclude that a good brake pad is one that is insensitive to (i.e., robust against) temperature, such as design D in Fig. 8.4. The braking force generated by brake pad D will vary little under different working temperatures. In other words, brake pad D will not generate excessive braking force at low temperatures, but will generate sufficient braking force at high temperatures. Design D is therefore insensitive to temperature. The friction coefficient of design D at low temperatures is lower than that of design C. In other words, the material of design D is not as soft as design C, making design D more resistant to wear and perhaps more durable than design C. In this example, brake pad temperature is a major noise factor. We expect the friction coefficient of brake pads to be insensitive to their temperature. If we can achieve this objective, the relationship between pedal force and braking force will be close to a linear proportionality and will have little variation, just like design D.

In this brake pad example, the brake system has only one basic function: to translate pedal force into braking force. To simultaneously solve various downstream quality problems such as braking efficiency, durability, squeaking, thermal radiation, etc., we need to improve the basic function of the brake system. Using traditional quality-engineering methods to solve downstream quality problems, we usually get a mediocre design, a trade-off among all downstream quality problems. Such a design seldom fundamentally solves all downstream quality problems. The most efficient way to fundamentally solve the quality problems of a system is to improve its basic function.

## CASE STUDY 3: DEVELOPMENT OF MACHINING TECHNOLOGY FOR HIGH-STRENGTH STEEL*

This case study involves the development of numerical control (NC) machining technology for high-strength steel. The basic concept of the study can be extended to the development of other manufacturing technologies such as injection molding, die casting, etc.

This case study was presented at the Tenth Taguchi Symposium, Costa Mesa, California in 1992 and received the Most Outstanding Case Study Award at the symposium. Its objective was to apply the concept of transformability (i.e., the capability to transform input signals into output responses) accurately and consistently to improve the basic functionality of NC machining technology. For this case study, we developed a generic test piece to evaluate the functional robustness of NC machining technology. The purpose was to show engineers how to formulate and conduct a robust engineering project. Below is a description of the problem-formulation procedure used in this case study.

To develop this machining technology, the author first discussed the project objective with the manager and engineers of the Machining Technology Department of his company. It was decided that the objective would be to use robust engineering methodologies to develop robust-machining technology for high-strength steel, which is very difficult to machine.

Although the engineers considered this NC machining process to be a generic machining technology, they had been using it to machine a very limited number of products (gears and ball-joint components). They were using two measurement characteristics, surface roughness and tool life, to evaluate the capability of their NC machining technology for specific products. The author considered these two measurement characteristics to be downstream quality characteristics not suited to the development of generic NC machining technology. One year prior to the project, these two downstream quality characteristics had been used several times to try to improve the accuracy of this NC machining technology, but there was no significant improvement. In the opinion of the author, using these two downstream quality characteristics as measurement characteristics might explain why they made so little progress in improving the accuracy of the NC machining technology

Before proceeding further, the basic function of the machining technology needed to be identified and stated. The materials of the target products (e.g., gears, ball-joint components) of the machining technology vary considerably in hardness. The basic function of this machining technology is to transform NC data on nominal product dimensions into actual product dimensions smoothly and accurately, regardless of hardness variations in raw materials. If this machining technology could predictably perform this basic function, its accuracy would be very high. Thus one critical issue in this robust engineering project was to select an appropriate measurement characteristic for the basic function of the machining technology.

The basic function of a product or process technology is related to its capability to transform input energy into output energy. If the energy-transformation efficiency is high and repeatable, the technology will be very robust in performing its basic function. The machining engineers involved in this case study believed it would be easy to measure the total output energy of the NC machining technology because considerable output energy is transformed into thermal friction energy. Since it is difficult to measure thermal friction energy precisely, we used another measurement characteristic to evaluate its basic function. This measurement characteristic is related to the dimensional transformability of the NC machining tech-nology, which is the capability of the machining technology to precisely translate dimensional input signals into corresponding output responses. The input signals in NC machining technology are numerical data describ-ing the nominal values of product dimensions, and the output responses are the actual dimensions of the machined products.

To ensure easy measurement of dimensional transformability, the test-piece shapes should not be too complicated. However, the ranges of input signals and output responses should be wide enough to cover the dimen-sions of possible future products. In addition, the test pieces should be three-dimensional (3-D) because future products are likely to be three-dimensional. Gears are not good test pieces because their shapes are too complicated to measure and they do not range widely in dimension. Further, it is difficult to measure gear dimensions because of their compli-cated three-dimensional curvatures. In short, using current products to develop a future generic technology will reduce the flexibility of the tech-nology. However, the functional robustness (i.e., dimensional transforma-bility) of the machining technology does need to be enhanced first. If the basic function of the machining technology is robust, we can use it to machine numerous future products (including gears and ball-joint compo-nents) smoothly and accurately. Since straight-line (i.e., one-dimensional) machining is the basis of all machining processes, the test pieces were composed of numerous straight lines. Two-dimensional surface machining and complicated three-dimensional machining are all based on one-dimen-sional straight-line machining. After this discussion, we developed a generic test piece, illustrated in Fig. 8.5, which is easy to measure and has a wide range of dimensions.

In order to meet strict requirements for strength and durability, gears (or ball-joint components) were generally processed using carbon-immersed heat treatment. Originally, the engineers used low-carbon steel as the raw material for gears in the NC machining technology described above. Low-carbon steel is soft, its hardness close to 0 on the Rockwell hardness scale. Thus, it is very easy to machine into complicated shapes like gears. However, after machining, it usually takes more than 10 hours to harden the machined gears using carbon-immersed heat treatments. Because the heat treatment process was not time efficient, productivity was usually very low. Even worse, gear shape and dimensions were usually distorted after the

**Current product**
**(e.g., gears)**          **Generic test piece**

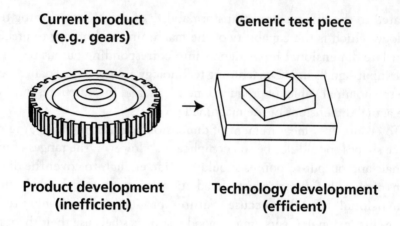

**Product development**      **Technology development**
**(inefficient)**                  **(efficient)**

**Figure 8.5** From product development to technology development.

long heat-treatment process. To increase productivity and reduce distortions in gear shape and dimensions, the machining engineers planned to harden gears using a high-frequency induction heat treatment process instead of the carbon-immersed heat treatment. This process takes only 1 minute to increase gear hardness. Thus, productivity would be increased and the dimensional distortion reduced. High-frequency induction heat treatment shows good potential to replace carbon-immersed heat treatment in the manufacturing technology bases of this company.

However, the high-frequency induction heat-treatment process can only harden high-strength steel (i.e., high-carbon steel), not low-carbon steel. The hardness of high-strength steel is around Rockwell 30, higher even than most low-carbon steel after carbon-immersed heat treatment. Because of its hardness, it is very difficult to machine into products with complicated shapes, such as gears or ball-joint components. Consequently, the machined surfaces of high-strength steel were usually rough, and the machining tools seldom lasted long. These are the two major reasons why the Machining Technology Department had to develop robust NC machining technology for high-strength steel.

Before applying robust engineering to develop this NC machining technology, the engineers focused exclusively on the machining of current products (e.g., gears). They applied company gears as test pieces to evaluate the capability of the NC machining technology to machine high-strength steel. Intuitively (but unfortunately), they used surface roughness of machined gears and tool life as the two capability measurement characteristics. In fact, both are downstream quality characteristics and not suited to measure the basic function of NC machining technology. As a result, the engineers did not improve machining accuracy or tool life despite tremendous effort. There are two major problems in using downstream quality characteristics to develop generic technologies:

1. Downstream quality characteristics are not as time efficient for developing robust technologies as upstream quality characteristics. It usually takes longer to use downstream quality characteristics to develop new technologies. Consequently, the company may lose its technological advantage.

2. Technologies developed through downstream quality characteristics are only good for the development of specific products and are not very flexible. Development engineers may need to conduct many adjustments of the newly developed technologies in the development of different products.

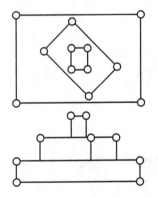

**Criteria for generic test pieces for NC machining technology development:**

1. Simple shapes for easy measurement

2. Wide range of input signals

3. Three-dimensional shape

**Figure 8.6** Test piece for NC machining-technology development.

At the technology-development stage, we should use generic test pieces, not specific products, to evaluate the functional robustness of the new technology. In this case study, machining engineers considered the size, shape, and dimensions of all possible future products and then developed the generic test piece illustrated in Fig. 8.6. Next, they used the NC machining technology and various high-strength-steel materials to make test pieces.

To enhance the flexibility of this machining technology, the engineers worked with the widest possible range of input signals (and thus output responses). After machining all test pieces, they used three-dimensional measurement equipment to measure the coordinates of 12 points on the test pieces, as illustrated in Fig. 8.6. The nominal values of the distances between these points are input signals, while the actual distances are output response, as illustrated in Fig. 8.7. Ideally, input signals and output responses should be linearly proportional to each other.

Each test piece has 12 points, for a total of 66 discrete distances between them. After the machining process, the engineers measured the (X, Y, Z) coordinates of the 12 points and calculated the distances between them using the measured data. The 66 actual distances were the output responses.

**Input signals: Nominal values of the 66 distances between the 12 points of each test piece**

**Output responses: The actual values for the 66 distances**

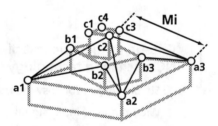

Number of input signals = number of output response levels = 66

**Figure 8.7** Input signals and output responses for test pieces.

The engineers used the proportionality between the input signals and output responses to describe the basic function (i.e., dimensional transformability) of the NC machining technology. Although ideally all input signals should be linearly proportional to output responses, because of material imperfections and various noise conditions in the NC machining process, the real function of this process showed considerable variation, as illustrated in Fig. 8.8.

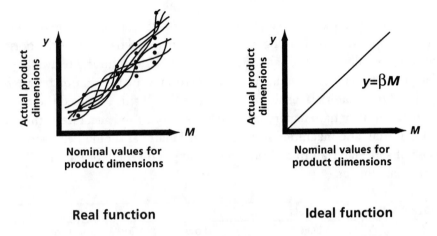

**Real function**                    **Ideal function**

**Figure 8.8** Basic function (dimensional transformability) of NC machining technology.

In this case study, the proportional constant ($\beta$) between input signals and output responses may not exactly equal 1. Simply put, the objective of this case study is to use robust engineering to first reduce the variation of this proportionality constant and then to adjust it to 1. If this proportionality constant is equal to 1 and has little variation, the output dimensions of products made by the NC machining process will be very close to the corresponding input signals (i.e., nominal values for product dimensions).

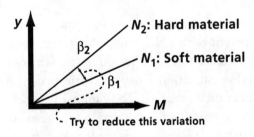

- Enhance linearity
- Reduce proportionality constant variation
- Improve functional robustness of the machining technology
- Downstream quality characteristics such as surface roughness and tool life will be improved accordingly

**Figure 8.9** Variation in the proportionality constant.

The proportionality constant should vary as little as possible for a wide range of input signals. The engineers selected material hardness as the noise factor to simulate functional variation of the NC machining technology. The proportionality constant was expected to show little variation between hard and soft materials, as illustrated in Fig. 8.9. If this is indeed the case, the machining of hard material should be as smooth and accurate as that of soft material, and machined surfaces of hard material should be as good as those of soft material. In addition, machining tools used for hard material should last longer.

**TABLE 8.1 Factors and levels**

| Factors | Levels | Low | Middle | High |
|---|---|---|---|---|
| | A: Machining direction | up | down | |
| | B: Machining speed | slow | standard | fast |
| | C: Feeding speed | slow | standard | fast |
| Control | D: Tool materials | soft | standard | hard |
| factors | E: Tool rigidity | low | standard | high |
| | F: Rotation angle | small | standard | large |
| | G: Contact angle | small | standard | large |
| | H: Cutting rate | small | standard | large |
| Noise factor | N: Material hardness | soft | hard | |

The engineers selected the control and noise factors listed in Table 8.1. During the development of this NC machining technology, they relied on their engineering judgment and experience to identify control factors that might affect its basic function. They also identified several noise factors that might affect functional variation. To simplify the experiment, the engineers applied only one noise factor—material hardness—to simulate variation effects. There were two noise factor levels: soft (class C-3 on the Rockwell hardness scale) and hard (class C+3).

Next, the engineers assigned the selected 8 control factors to an $L_{18}$ inner orthogonal array. They then assigned the two-level noise factor (material hardness) to every row of the array. This produced 36 experimental runs in total (and thus 36 test pieces). They randomized all 36 experiments and then conducted them. They produced one S/N ratio (i.e., robustness measurement) and one sensitivity S (i.e., proportionality constant after logarithm transformation) for each row of the array, as shown in Table 8.2. Since the focus of this book is not the analytical details of robust engineering, the author will not illustrate calculations on the S/N ratio and sensitivity. Please refer to *Introduction to Quality Engineering*, Volume 1, American Supplier Institute (ASI) or the *Proceedings of the Tenth Taguchi Symposium* (ASI), for more details on calculations of S/N ratio and sensitivity S.

**TABLE 8.2 Experimental layout of the $L_{18}$ inner orthogonal array and the analytical results**

|    | A | B | C | D | E | F | G | H | S/N ratio (dB) | Sensitivity |
|----|---|---|---|---|---|---|---|---|----------------|-------------|
| 1  | 1 | 1 | 1 | 1 | 1 | 1 | 1 | 1 | 31.41 | -0.0022 |
| 2  | 1 | 1 | 2 | 2 | 2 | 2 | 2 | 2 | 39.70 | 0.0058 |
| 3  | 1 | 1 | 3 | 3 | 3 | 3 | 3 | 3 | 39.68 | 0.0028 |
| 4  | 1 | 2 | 1 | 1 | 2 | 2 | 3 | 3 | 9.25 | 0.0730 |
| 5  | 1 | 2 | 2 | 2 | 3 | 3 | 1 | 1 | 44.56 | -0.0001 |
| 6  | 1 | 2 | 3 | 3 | 1 | 1 | 2 | 2 | 42.02 | 0.0020 |
| 7  | 1 | 3 | 1 | 2 | 1 | 3 | 2 | 3 | 33.75 | 0.0057 |
| 8  | 1 | 3 | 2 | 3 | 2 | 1 | 3 | 1 | 44.59 | 0.0003 |
| 9  | 1 | 3 | 3 | 1 | 3 | 2 | 1 | 2 | 19.18 | 0.0114 |
| 10 | 2 | 1 | 1 | 3 | 3 | 2 | 2 | 1 | 42.80 | 0.0011 |
| 11 | 2 | 1 | 2 | 1 | 1 | 3 | 3 | 2 | 30.55 | 0.0145 |
| 12 | 2 | 1 | 3 | 2 | 2 | 1 | 1 | 3 | 26.41 | 0.0166 |
| 13 | 2 | 2 | 1 | 2 | 3 | 1 | 3 | 2 | 25.86 | 0.0148 |
| 14 | 2 | 2 | 2 | 3 | 1 | 2 | 1 | 3 | 35.24 | 0.0056 |
| 15 | 2 | 2 | 3 | 1 | 2 | 3 | 2 | 1 | 42.52 | 0.0022 |
| 16 | 2 | 3 | 1 | 3 | 2 | 3 | 1 | 2 | 41.01 | -0.0009 |
| 17 | 2 | 3 | 2 | 1 | 3 | 1 | 2 | 3 | 2.63 | 0.1801 |
| 18 | 2 | 3 | 3 | 2 | 1 | 2 | 3 | 1 | 39.30 | 0.0025 |

**TABLE 8.3 Optimal and initial conditions**

| Optimal condition: | $A_1B_1C_3D_3E_1F_3G_2H_1$ | |
|---|---|---|
| Initial condition: | $A_1B_2C_2D_2E_2F_2G_2H_2$ | |
|  | S/N ratio (dB) | Proportionality ($\beta$) |
| Optimal condition | 57.24 | 0.9935 |
| Initial condition | 33.73 | 0.9989 |
| Gain | 23.51 | |

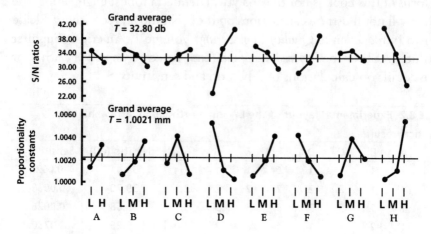

**Figure 8.10** Main-effect plots for control factors for S/N ratios and proportionality constants.

Using the test results, the engineers were able to calculate the main effects of the 8 control factors on S/N ratios and proportionality constants, as illustrated in Fig. 8.10.

The engineers selected the most significant control factors on S/N ratios from the main-effect plots in Fig. 8.10 so as to maximize the functional robustness of this machining technology. Sensitivity S is the logarithm transformation of the proportionality constant ($\beta$).

The objective of this case study was to improve the functional robustness of NC machining technology for high-strength steel. To achieve this objective, we needed to maximize the S/N ratio of the technology's basic function (i.e., transformability from nominal values of product dimensions into actual product dimensions). The optimal condition (i.e., the condition that maximizes functional robustness) and initial condition of this machining technology are $(A_1B_1C_3D_3E_1F_3G_2H_1)$ and $(A_1B_2C_2D_2E_2F_2G_2H_2)$, respectively.

Finally, the engineers used the gain in S/N ratio to estimate improvement in functional robustness between optimal and initial conditions, as shown in Table 8.3.

The S/N ratio of the optimal condition in this case study is 23.5 dB higher than that of the initial condition. This improvement means that the variation of the optimal condition is only 1/220 of the variation of the initial condition. In other words, standard deviation of the optimal condition is only 1/15 that of the initial condition. Therefore the accuracy of the machining technology was improved dramatically. The objective of this case study was to maximize functional robustness. It is not necessary to adjust the proportionality constant to exactly 1. We can adjust the input signal M to $y_0/\beta$ to ensure that output product dimensions $y_0$ meet nominal targets. In Table 8.3, the proportionality constant ($\beta$) of the optimal condition was slightly smaller than that of the initial condition, but this was not really a problem.

Next, the engineers conducted confirmation tests for both optimal and initial conditions. The results are listed in Table 8.4, from which we can see that the gain in S/N ratio between optimal and initial conditions was confirmed. The proportionality constants in the confirmation test were about the same as in the original tests. The gain in S/N ratio produced in the confirmation tests is 19.4 dB, close to 23.5 dB in the original tests. In the opinion of the author, the difference between the two results is due to hardness variation in the test pieces. The confirmation tests also confirmed that the surface roughness under optimal conditions and machining tool life improved over the initial condition.

**TABLE 8.4 Comparison between original tests and confirmation tests**

|  | Original tests | | | Confirmation tests | | |
|---|---|---|---|---|---|---|
|  | Optimal | Initial | Gain | Optimal | Initial | Gain |
| S/N ratio | 57.24 | 33.73 | 23.51 | 54.09 | 34.71 | 19.38 |
| $\beta$ | 0.9935 | 0.9989 | | 0.9939 | 0.9992 | |

When the author presented this case study at the Tenth Taguchi Symposium, several questions were raised by the audience. One is related to S/N ratios on the $L_{18}$ orthogonal array shown in Table 8.2. The S/N ratios range between 44.59 dB (Row No. 8) and 2.63 dB (Row No. 17). The average of these 18 S/N ratios is 32.8 dB. The S/N ratios of Rows 4 and 17 are 9.25 and 2.53 dB, respectively, much lower than the average value, 32.8 dB. Some people thought that there might be something wrong with these two S/N ratios.

In this case study, the range of control-factor levels was great and the noise factor (material hardness) also varied significantly. Even worse, machining oil was not used to lubricate and cool machining tools during the experiments, so some machining conditions were really bad. In the experiments in Rows 4 and 17, scrap metal melted onto the machining tools and the machined surfaces of these two experiments were very rough. It is therefore not surprising that the S/N ratios of Rows 4 and 17 experiments were very low.

On the other hand, the finished surfaces of experiments that had high S/N ratios are all smooth. In addition, the machining tools in those experiments also lasted longer than those with low S/N ratios. Therefore it was concluded that there was nothing wrong with the S/N ratios in Rows 4 and 17.

Another person questioned the repeatability of this NC machining technology in manufacturing specific products, as the purpose of this case study was to develop new machining technology, not specific products. If the engineers applied this machining technology to manufacture a specific product, they might not get the same results as shown in Table 8.2. The author disagreed. This NC machining technology had input signals (i.e., nominal values for product dimensions) in X, Y, and Z directions independently. The output responses (i.e., actual product dimensions) should be as close to the associated input signals as possible. If the proportionality constant between input signals and output responses varies little, the dimensions of the machined test pieces or specific products will be very close to the specifications (input signals). Thus, the shapes of the machined products will be very close to original specifications. If the NC machining technology is of high accuracy, it can be applied to machine products of any shape accurately and predictably. In other words, the transformability of this machining technology is very predictable over a wide range of potential future products.

Another participant commented on the proportionality constant between input signals and output responses, stating that it was not necessary to calibrate control factors to adjust the proportionality constant to 1 because machining engineers can directly adjust the input signals to control output responses. The author agreed. It is not really necessary to make the proportionality constant equal 1. Machining engineers can always calibrate the input signal instead of the proportionality constant to produce output product dimensions that are very close to nominal values. For instance, assume that the nominal value of a product dimension is $y_0$ and the proportionality constant is already known to be $\beta$. Machining engineers can set the input signal to $M = y_0/\beta$. This way, they can get an output response of $y_0$, if the proportionality constant has little variation.

At the specific product-development stage, the absolute value of $\beta$ may be crucial to machining accuracy. However, at the development stage of a generic technology, the functional robustness of the generic technology is, in fact, more important than the absolute value of $\beta$. Thus, at the technology-development stage, development engineers need to focus on reducing the variation of the functionality of this technology. If product-development engineers apply the optimized technology to machine specific products such as gears, the value of $\beta$ might be slightly different from that of the original tests shown in Fig. 8.6. However, it is not difficult to calibrate control factors to adjust the absolute value of $\beta$ in order to improve machining accuracy for specific products. In short, at the technology-development stage, the most critical issue is to find a good combination of control factors to enhance the functional robustness of the target technology, not to calibrate the absolute values of the proportionality constant.

# Implementations of Robust-Technology Development

## 9.1    NEW TECHNOLOGIES AND AUTOMOBILE INDUSTRIES

An automobile is composed of numerous systems, such as the engine, transmission, suspension, and so on. The basic mechanical structures of these systems have not significantly changed since their invention and application in the automobile industry. Thus, these systems are easily understandable for most engineers. The author would first like to illustrate how the advancement of new technologies affects the automobile industry.

One example of an automotive technology is front-suspension technology. Major automobile manufacturers first used a strut-type front suspension during the period between the early 1970s and the first half of the 1980s. The technology was developed and patented by BMW Co. of Germany during the 1960s. Before 1970, major automobile manufacturers used a wishbone-type front-suspension technology in their automobiles. However, a strut-type front suspension is less complicated and cheaper than the wishbone-type, but retains all its advantages, such as high-speed stability, low noise, structural integrity, etc. Thus, strut-type suspension technology replaced wishbone-suspension technology in the automotive mainstream. However, because of increased requirements associated with high-speed performance, many automotive companies tended to return to a wishbone-type front suspension to reduce noise and airflow resistance and to increase suspension rigidity. As we can see from this example, technology is constantly changing, even for such a subsystem as front suspension. Consequently, it is very important to select appropriate technologies for the development of new products since both quality and performance will be significantly affected at the upstream stage of product planning.

A second example is the use of electronic control devices for such automotive systems as emission control, air/fuel control, antilock brake system (ABS), and so on. As mentioned in Chapter 3, electronic control systems are finding their way into automobiles at an unbelievable rate. Most are built upon basic mechanical structures, which evolved over a long period. In order to meet both high-performance requirements and strict legislation, many sophisticated electronic control systems were added to the basic mechanical structures. However, if these structures are not inherently robust, the corresponding output responses may experience considerable variation, and the electronic control devices may need to work actively to compensate for the performance variation. Thus, the electronic control devices may not be able to last very long. In most automotive applications, the addition of electronic control devices does not fundamentally solve quality problems if the basic mechanical structures are not functionally robust.

In this high-tech era, the fundamental way to solve quality problems is to apply robust engineering to develop robust technologies for future products. Take the evolution of front-suspension technologies, for example. There are numerous front-suspension technologies: strut type, wishbone type, trailing-arm type, multilink type, etc. Each technology has its own basic function. To enhance the functional robustness of the target technology, we need to both develop a measurement for its robustness and enhance its functionality (i.e., efficiency in translating input signals into output responses). In other words, we need to reduce the functional variation (i.e., instability or nonrobustness) of the target technology. If we can reduce functional variation, we can reduce cost by lowering the average performance-characteristic values to just above specification limits, as illustrated in Fig. 9.1. In this way, we will be able to reduce the size, weight, and cost of the new technology. If we can apply robust engineering to develop reliable and stable technologies, we will have more opportunity to enhance the quality of new products while reducing average development cost. On the other hand, if the functional variation of the new system is large, we need to raise the mean value of the quality or performance characteristic above specification limits to ensure that all mass-produced products meet the performance specification, as shown in Fig. 9.1. As a result, the average cost of the product will be very high. If we can ensure the functional robustness of a new technology at its early development stage, we will be able to save a lot of engineering resources in later stages. Thus, at the technology-development stage, we need to focus on reducing the functional variation of the new technology instead of adjusting quality or performance characteristics to meet design specifications. Such adjustments should be conducted after robust technologies are developed.

We expect to reduce the functional variation of a new technology at the very early development stage. If the requirements of the quality or performance characteristics are specified at later stages, engineers can adjust the mean value to meet the specification limits without overengineering. In this way, engineers can reduce the weight and manufacturing cost of new products and can even calibrate the performance characteristics of the newly developed system to meet numerous customer requirements. Certainly,

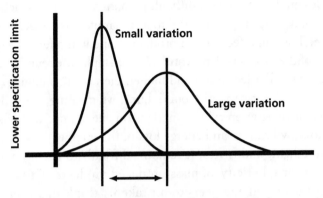

**Figure 9.1** Specification limit and mean value for performance (or quality) characteristic.

numerous electronic control devices and system technologies for the new technology must be developed and made robust at the same time.

After new products are planned and performance requirements specified, we can adjust the generic product technology to meet numerous targets by considering the requirements of performance, weight, and cost. After product planning and marketing, we can still conduct further fine-tuning activities to optimize performance or quality characteristics and reduce the cost of the newly developed products even further. If the basic functions of the newly developed products are already robust, fine-tuning will be very easy. The new products can thereby be delivered to market in a very time-efficient way.

## 9.2   MANUFACTURING INDUSTRIES AND ROBUST TECHNOLOGIES

Many product quality problems are actually due to imperfections or nonro-bustness in the associated manufacturing technologies. In manufacturing industries, many quality problems cannot be solved through design activities. For example, the roundness imperfections in cylinders or pistons are due to the inaccuracy of the associated machining technologies and have nothing to do with design. It is extremely difficult to manufacture a product of perfect shape (e.g., roundness or sphericity) if robust manufacturing technologies are not in place. The imperfections in product shape will affect the quality or performance of the products significantly. For example, the roundness of cylinders and pistons will affect noise and fuel efficiency in an engine. A piston or cylinder may appear perfectly round, but if we measure, we will find it is not. The clearance between cylinder and piston due to imperfections in shape will cause noise, vibration, and energy loss in the engine.

In manufacturing industries, it remains difficult to accurately measure the roundness, or sphericity, of mass-produced products. This difficulty in measurement may lead engineers to mistakenly think that their products are perfect in shape, and they may end up misjudging the quality or performance of the products. If they can precisely measure product shape and apply robust-manufacturing technologies to reduce variation in shape, they will be able to efficiently improve the associated downstream quality characteristics (e.g., engine noise) .

In the case study in Chapter 8, we also discussed the shape (i.e., sphericity) imperfections in ball-joint components. This quality problem could not be fundamentally solved through design activities. The key in this case study was to make the socket cavities and the associated ballstud heads perfectly spherical through NC manufacturing technology of high accuracy and repeatability. Simply put, this quality problem was related to manufacturing technologies, not design activities. It was difficult to measure the sphericity of socket cavities and ballstud heads, but to reduce friction in ball joints, we needed to machine the cavities and ballstud heads to their most spherical shape through robust NC machining technologies.

The capability to develop robust technologies in a manufacturing company is critical to its competitiveness, as any manufacturing company with a base of advanced and robust-manufacturing technologies can make exclusive products with complicated shapes or sophisticated performance characteristics, products that are so difficult to measure that other companies simply cannot compete. In comparison, companies that do not have such a base can only make noncompetitive products with easy-to-measure shapes or performance characteristics. A manufacturing company without robust-manufacturing technologies cannot fundamentally solve downstream quality problems, because such quality problems are usually due to the instability of the associated manufacturing technologies. It takes a long time to develop such a base of robust-manufacturing technology, but such a base is the key to competitiveness for any manufacturing company.

Robust engineering plays a decisive role in the development of manufacturing technologies. The concepts of transformability (i.e., transforming input signals into output responses) in dynamic-type robust engineering, a recent development, is especially critical to the development of robust-manufacturing technologies. Downstream measurement or inspection activities, though still important to manufacturing industries, are not efficient ways to fundamentally solve quality problems. In many industrial applications, it remains difficult to accurately measure quality characteristics (e.g., cylinder or piston roundness) or performance characteristics of manufactured products. The most fundamental way to solve manufacturing quality problems is to develop robust-manufacturing technologies. High-quality products can be manufactured and delivered to market in a time-efficient way through a base of such technologies. Downstream measurement and inspection activities cannot solve manufacturing problems time efficiently.

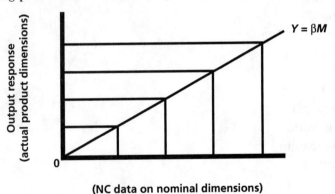

**Figure 9.2** Proportionality relationship between input signals (NC data on nominal dimensions) and output responses (actual product dimensions).

In the case study in Chapter 8, we developed robust NC machining technology to solve the quality problem of sphericity imperfections in ballstud heads and socket cavities for ball joints. Let the nominal values for product dimensions be input signals (along X, Y, and Z axes, respectively) and the corresponding dimensions of the products processed by the NC technology the output responses. Ideally, there should be a linear proportionality relationship between the two. However, in critical industrial applications, this relationship may be more complicated. The objective of this robust engineering case study is to make real function approach linear proportionality, as illustrated in Fig. 9.2. If real function of the NC machining technology is close to linear proportionality under various operating conditions, it will be inherently robust. If we then use this NC machining technology to machine the ballstud heads or the socket cavities of ball joints, their shapes will be of nearly perfect sphericity. As a result, the assembled ball joints should experience little friction force.

In the opinion of the author, the same robust engineering approach seen in Case Study 3 in Chapter 8 can be applied to the following manufacturing technologies. If robust engineering can be successfully applied to improve functional robustness, their productivity will increase dramatically.

- Injection molding
- Vacuum forming
- Gum forming
- Die casting
- Forging
- IC manufacturing
- NC machining
- Drilling
- Stamping
- Photocopying
- Photographing
- Video recording
- Facsimile
- X-ray
- and others

The dynamic-type of robust engineering illustrated in the case study in Chapter 8 can be used to similarly improve the functional robustness of the above technologies. Dynamic-type robust engineering has been proven efficient in improving the functional robustness of welding and material technologies at the author's company. Robust engineering is indeed the deciding factor in the efficient development of new technologies.

If one manufacturing company can develop robust technologies and apply them to develop new products, its products will be the most competitive, both in terms of quality and manufacturing cost. It is essential to company competitiveness to develop robust technologies to ensure the technical capability to design and manufacture high-quality products at low cost. Product and process technologies therefore need to be developed even before design activities for any specific product begin. At the product-development stage, all key product and process technologies should already be in place, so that new products are ready to be mass produced immediately after being planned and designed. If a manufacturing company can develop robust and flexible technologies, it will be able to put newly developed products into mass production in a time-efficient way by simply calibrating adjustment factors.

Product planning is related to marketing activities for new products. It usually takes time to conduct product planning or marketing, after which not much time is left for new-product development if key technologies are not ready. To reduce time to market, engineers need to develop robust-product technologies and associated process technologies before product planning or marketing. Bringing robust engineering to technology development is a new and efficient way to reduce time to market. After robust product and process technologies are developed, engineers can conduct product design, prototyping, and validation testing to develop new products immediately after product planning and marketing. Of course, process technologies are as important as product technologies. A manufacturing company needs to develop robust-manufacturing technologies to ensure their ability to manufacture newly developed products. Only through the development of robust product and process technologies can a manufacturing company deliver high-quality products at low cost in a time-efficient way, thereby gaining an advantage over the competition.

To sum up, developing robust product and process technologies is the most efficient way to improve the productivity of a manufacturing company. The ability to develop robust technologies will be the most important factor in competitiveness in the near future. To improve the efficiency of the total-product-development process, a manufacturing company needs to invest a tremendous amount of engineering resources in developing a new generation of product or process technologies. Efficiency in the development of robust technologies is the deciding factor in a manufacturing company's productivity.

## CASE STUDY 4: DEVELOPMENT OF CFRP (CARBON-FIBER-REINFORCED PLASTIC) INJECTING-MOLDING TECHNOLOGY*

Many automobile components are made from plastic materials. These account for about 8% of the weight of an automobile. The major reason for using plastics in automobiles is to reduce the total weight so as to improve fuel economy and driveability. It is common to manufacture plastic components with complicated shapes using injection-molding processes. Because of their high operating efficiency and flexible shaping capability, injection-molding technologies have tended to replace stamping technologies in industrial applications. In other words, they are more productive than stamping technologies.

However, plastic materials are usually weaker than metal materials. Thus, in many applications, plastic components must be thicker to ensure strength and functionality. As a result, the plastic components may become heavy, which defeats the reasons for using them in the first place. In addition, increased thickness may reduce the dimensional accuracy of the components. Low dimensional accuracy is a significant problem in typical plastic-injection-molding processes. There are three common quality problems involved in plastic-injection-molding processes:

- Dimensional variation and shape distortion
- Internal shrinkage
- Shrinkage or expansion of external surface

Repeated trials are commonly required to adjust the dimensions of injection molds or the settings of an injection-molding process to produce acceptable plastic components, as it is difficult to inject high-quality plastic components consistently.

Development engineers in the plastic department wanted to apply robust engineering to develop a plastic-injection-molding technology for CFRP (carbon-fiber-reinforced plastic) materials, which are stronger than ordinary plastic materials. The objective of this project was to determine the optimal operating condition of the CFRP injection-molding technology to prevent the internal shrinkage that occurs in this type of material. Prior to this robust engineering project, the engineers had already conducted several preliminary experiments to inject certain CFRP components and had accumulated enough technological knowledge to produce CFRP components of middle quality. However, they were unable to produce high-quality CFRP components consistently and predictably. Thus, they needed to develop a robust injection-molding technology for CFRP materials.

Initially, the engineers wanted to use CFRP material to develop rotors (i.e., turbine rotors) for turbo-chargers in high-performance gasoline engines. Such rotors are usually made of metal materials, but if they could use CFRP materials with high dimensional accuracy, they would be able to reduce the weight of the rotors and thus improve the performance of turbo-charged engines. A turbine rotor has many blades, the shapes of which are usually complicated. The rotor's working temperature is usually over 200°C, too high for ordinary plastic materials. CFRP materials are a mixture of super-engineered plastic and various types of carbon fiber and can be used to make plastic components of complicated shapes using injection-molding technologies. The melting point of CFRP material is much higher than that of ordinary plastic materials; however, if the working temperature is 50° higher than the material's melting point, the plastic may be carbonized and lose strength and other performance characteristics.

When we conducted robust engineering to improve the functional robustness of the CFRP injection-molding technology, we refrained from focusing on a current product or specific component, such as an automobile turbine rotor. Instead, we concentrated on developing a generic test piece and simulating various noise conditions to estimate the functional variation of the technology. We then adjusted the various control factors of the technology to determine the optimal operating conditions so as to minimize functional variation.

Prior to this project, these engineers had already determined certain operating conditions that could prevent the occurrence of internal shrinkage. However, the conditions were not easily repeatable, and visible defects caused by internal shrinkage still occurred occasionally. Therefore, they wanted to use dynamic-type robust engineering to enhance the robustness of the CFRP injection-molding technology. They also wanted to determine what component shapes were appropriate for this technology.

At the very beginning of this project, the engineers used an automobile turbine rotor, a current product, as the test piece to conduct experiments, and then measured its key dimensions. It had numerous blades of about 60-mm outside diameter and 0.6 to 0.8-mm in front-blade thickness. They also measured shape distortion of the rotor and the corresponding dimensions of the injection mold. They found that it was very difficult to measure blade surface curvatures and shapes. In fact, they were unable to determine a reliable measurement technique for these dimensions and were thus unable to use dynamic-type robust engineering methods and the concepts of transformability illustrated in Fig. 9.2 to improve the functional robustness of the injection-molding technology. The engineers and the author discussed how to go about initiating this robust engineering project. Per the author's request, the engineers developed a generic test piece, as illustrated in Fig. 9.4, on which to conduct the experiments, replacing the original plastic product, the rotor shown in Fig. 9.3. At the technology-development stage, we should disregard current products and instead focus our resources on the development of generic test pieces so as to enhance the flexibility of the target technology.

The purpose of the functional transformability shown in Fig. 9.2 is to make output responses be linearly proportional to corresponding input signals. In the case study, we expected the actual product dimensions (output responses) to be linearly proportional to the corresponding dimensions (input signals) of the injection mold. If the relationship between input signals and output responses approaches linear proportionality under various operating conditions, the basic function of the technology will be very robust. As a result, the dimensions of injected test pieces should be linearly proportional to the dimensions of injection molds. If the basic function of this injection-molding technology is not robust, the mold dimensions will need extensive calibrations to compensate for the functional instability of the technology. This calibration process is very time consuming for those working with injection molds. The most important issue in this case study was to make the output-product dimensions linearly proportional to the mold dimensions, not to reduce the difference between product dimensions and the nominal dimensions of design drawings. The ratio between actual product dimensions and corresponding mold dimensions is defined as proportionality (i.e., $\beta$ or shrinkage ratio). We expect this shrinkage ratio to remain constant under all noise conditions. If the shrinkage ratio remains constant, it will be easy to calibrate mold dimensions to make the corresponding product dimensions meet the nominal values of design drawings.

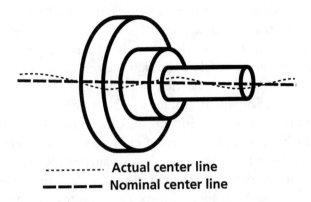

----------- **Actual center line**
━ ━ ━ ━ **Nominal center line**

**Figure 9.3** A current product (appropriate for product development, but not technology development).

The robust engineering approach for this injection-molding technology is very similar to the case study involving NC machining-technology development in Chapter 8. The purpose of both projects was to develop generic and robust technologies for future products, not for current products. If a manufacturing company does not develop a base of robust technologies, it cannot deliver future products to market in a time-efficient way. Without robust technologies ready and in place, development engineers will have difficulties in developing future products or putting future products into mass production. In this case study, we first developed a generic test piece for the measurement of the functional robustness of the injection-molding technology. Easy measurement is a critical factor in the design of the generic test piece. This piece is three-dimensional and very similar to the generic test piece for the NC machining technology in Case Study 3. Next, we measured the dimensions of the injection mold and the corresponding dimensions of the injected test pieces. Mold dimensions should be linearly proportional to product dimensions.

Some common products made of CFRP materials are rotary automotive components, such as turbine rotors, which usually rotate at very high speed. Therefore, eccentricity of these components is a critical quality characteristic. Figure 9.3 shows a rotor made of CFRP material, a rotor composed of three cylinders of different thicknesses and diameters. Theoretically, the center line of these three cylinders should all coincide. However, one measurement problem is the difficulty in determining this true center line. One engineer suggested that a line connecting the center point of the top surface and that of the bottom surface could be treated as the center line of the rotor, but there was some skepticism. Another measurement problem was that the top surface of the first cylinder and the bottom surface of the third cylinder are not perfectly parallel. These two measurement problems make it difficult to determine the true center line of this rotor. In fact, the center line of the rotor may not be a straight line, or it may not even exist at all. It was very difficult to accurately measure the dimensions (e.g., diameters, lengths, center line, thicknesses) of the rotor, even though it appeared straightforward. Another problem with using this rotor as the test piece was that it did not represent all possible products involving CFRP injection-molding technology. For the above reasons, we decided not to use this rotor to develop the CFRP injection-molding technology.

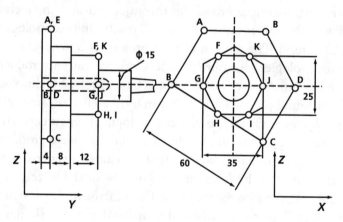

**Figure 9.4** Generic test piece for the development of CFRP injection-molding technology.

Accurate measurement is critical for the development of robust technologies, and therefore, we need to develop reliable measurement techniques for test pieces before conducting robust engineering activities to improve functional robustness. To make measurements easy and accurate, we developed the generic test piece illustrated in Fig. 9.4. A new challenge surfaced: how to measure the coordinates of the edge points (A, B, C, ... K) of the test piece. In the design drawing, all these edge points are simple points. However, after the test piece was injected, we found that these edges were not simple points, but rather very small three-dimensional curvature surfaces, as illustrated in Fig. 9.5. Thus, it was not as easy as it appeared to measure the coordinates of these edges. One engineer suggested that we could first determine the coordinates of the four side lines (a', a'', a''', a''''), which are parallel to the edge lines at a distance of 1 mm. We measured the coordinates of lines a', a'', a''', and a'''', and then calculated the interaction points (A', A'', A''') using geometrical equations. Finally, using a computer, we calculated the coordinates of the edge point A through the least-squared-error method. Obviously, this measurement technique and the associated calculation are very time consuming. However, this was the only way for accurate measurement of these edge points. This example illustrates the difficulties of industrial measurement applications for which engineering students are not prepared.

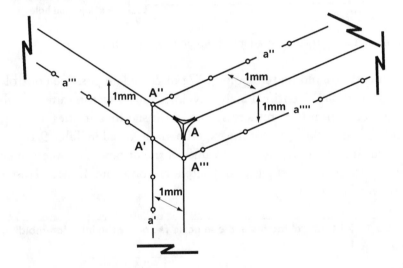

**Figure 9.5** Edge-point measurement.

For the generic test piece in Fig. 9.4, the section thickness increases from the bottom (ABCDE) to the top (cylinder = 15 mm in diameter), while areas increase from top to bottom. Using the injection-molding machine shown in Fig. 9.6, the heated CFRP materials were injected from the top to the bottom section and then cooled inside the mold. Thus the cooling process affected the thickness homogeneity and dimensional accuracy of the four sections of the injected test piece. The distance between injection nozzle and the mold was also an important factor for the functional robustness of

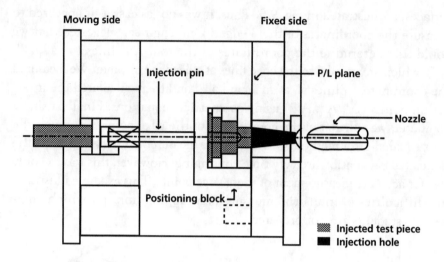

**Figure 9.6** Basic structure of an injection-molding machine.

this injection-molding technology. We discussed all possible control and noise factors, and identified many possible control factors, after which we conducted preliminary tests to screen out insignificant ones. Finally, we selected 8 control factors for the experiment, as listed in Table 9.1.

Input signals are the nominal values of the distances between the edge points of the generic test piece, and output responses are the actual distances between these points.

**TABLE 9.1 Eight control factors and one noise factor for the injection-molding technology**

| Factors | Levels | 1 | 2 | 3 |
|---|---|---|---|---|
| | A: Cylinder temperature (°C) | *low | high | |
| | B: Mold temperature (°C) | low | *middle | high |
| Control | C: Nozzle temperature (°C) | low | *middle | high |
| factors | D: Injection speed (%) | 5 | *8 | 15 |
| | E: Pressure holding position | 5 | *10 | 15 |
| | F: Holding pressure (%) | 70 | *85 | 99 |
| | G: Holding time (seconds) | 15 | *30 | 45 |
| | H: Cooling time (seconds) | 40 | *110 | 180 |
| Noise factor | N: Shot sequence number | 2nd shot | 5th shot | |

*Initial condition

In addition to the eight control factors, we also selected one noise factor to estimate the variation effect of this injection-molding machine. This noise factor is the shot sequence number of injected test pieces. We injected five plastic test pieces for each combination of the eight control factors and used the second and fifth shots for measurement. The reason for selecting shot sequence number as the noise factor was to ensure that the injection-molding process stabilizes as quickly as possible. Commonly, an injection-molding machine may need to inject more than 50 or even 100 defective shots before it stabilizes. One purpose of robust engineering is to reduce development cost. We expected the injection molding process to stabilize very quickly and the dimensions of injected components to be as accurate as possible. For these reasons we selected shot sequence number as the noise factor to measure the stability of the injection-molding technology. Generally speaking, we expect this injection-molding process to stabilize within 10 shots.

All eight control factors were assigned to an $L_{18}$ orthogonal array. In this experiment, we injected five shots for each row of the $L_{18}$ orthogonal array and measured the test pieces of the second and the fifth shots. After measuring the coordinates of the edge points of the test pieces, we calculated the sensitivities (S) and the S/N ratios of these 18 rows, as illustrated in Table 9.2. The main-effect plots of S/N ratios and sensitivities are illustrated in Fig. 9.7. In fact, sensitivity was related to the shrinkage ratio of the CFRP material. We can convert sensitivity values into the proportionality constant ($\beta$) between input signals (mold dimensions) and output responses (dimensions of injected test pieces).

The shrinkage ratios of CFRP materials are very sensitive to the operating conditions of the injection-molding machine. We needed to investigate all possible factors that could affect the proportionality relationship between mold dimensions and corresponding product dimensions. The goal of robust engineering is to determine control factors to maximize S/N ratios and thus the stability (i.e., robustness) of this injection-molding technology. From the main-effect plots of S/N ratios shown in Fig. 9.7, we see that the S/N ratio plots of factors E and H are both V shapes. This indicates that these two factors have nonlinear effects on S/N ratios. Robust engineering must take advantage of the nonlinear effects of control factors on S/N ratios to achieve robustness efficiently. The second level (the initial setting) of both factors is lower than levels 1 and 3. This indicates that we will be able to make improvement by switching from level 2 (initial

**TABLE 9.2 S/N ratios and sensitivity data (unit = dB)**

|    | A | B | C | D | E | F | G | H | S/N ratio $\eta$ | Sensitivity S |
|----|---|---|---|---|---|---|---|---|------------------|---------------|
| 1  | 1 | 1 | 1 | 1 | 1 | 1 | 1 | 1 | 16.70 | -0.0569 |
| 2  | 1 | 1 | 2 | 2 | 2 | 2 | 2 | 2 | 16.96 | -0.0557 |
| 3  | 1 | 1 | 3 | 3 | 3 | 3 | 3 | 3 | 17.23 | -0.0531 |
| 4  | 1 | 2 | 1 | 1 | 2 | 2 | 3 | 3 | 17.00 | -0.0550 |
| 5  | 1 | 2 | 2 | 2 | 3 | 3 | 1 | 1 | 16.46 | -0.0575 |
| 6  | 1 | 2 | 3 | 3 | 1 | 1 | 2 | 2 | 15.83 | -0.0592 |
| 7  | 1 | 3 | 1 | 2 | 1 | 3 | 2 | 3 | 16.04 | -0.0561 |
| 8  | 1 | 3 | 2 | 3 | 2 | 1 | 3 | 1 | 13.97 | -0.0639 |
| 9  | 1 | 3 | 3 | 1 | 3 | 2 | 1 | 2 | 14.46 | -0.0686 |
| 10 | 2 | 1 | 1 | 3 | 3 | 2 | 2 | 1 | 17.48 | -0.0489 |
| 11 | 2 | 1 | 2 | 1 | 1 | 3 | 3 | 2 | 18.17 | -0.0454 |
| 12 | 2 | 1 | 3 | 2 | 2 | 1 | 1 | 3 | 17.01 | -0.0541 |
| 13 | 2 | 2 | 1 | 2 | 3 | 1 | 3 | 2 | 16.86 | -0.0498 |
| 14 | 2 | 2 | 2 | 3 | 1 | 2 | 1 | 3 | 14.91 | -0.0612 |
| 15 | 2 | 2 | 3 | 1 | 2 | 3 | 2 | 1 | 17.87 | -0.0406 |
| 16 | 2 | 3 | 1 | 3 | 2 | 3 | 1 | 2 | 12.67 | -0.0782 |
| 17 | 2 | 3 | 2 | 1 | 3 | 1 | 2 | 3 | 16.24 | -0.0492 |
| 18 | 2 | 3 | 3 | 2 | 1 | 2 | 3 | 1 | 16.90 | -0.0449 |

setting) to level 1 or 3. If the main-effect plot of S/N ratio is a reversed-V shape, level 2 will be better than levels 1 and 3. In both the V shape or reversed-V shape plots of S/N ratios, there are significant interactions between control factors and noise factors. If the main-effect plots of some factors are V shape or reversed-V shape, we need to conduct validation tests to see whether or not they are reproducible.

The goal of robust engineering is to desensitize target technology (or a product or process) against noise factors through interactions between control and noise factors. Without robust engineering, products (or technologies) may perform unpredictably or inconsistently under customer-usage conditions, and many downstream quality problems may occur accordingly. If there are no significant interactions between control and noise factors, we may not be able to achieve robustness at low cost.

In this case study, the optimal condition that maximizes the S/N ratio is $(A_2, B_1, C_3, D_1, E_3, F_3, G_2, H_1)$. However, it was not easy to recalibrate factor C (the temperature of injection nozzle) from the initial setting (level 2) to level 3. Thus, the final setting was $(A_2, B_1, C_2, D_1, E_3, F_3, G_2, H_1)$. In comparison, the initial control factor settings were $(A_1, B_2, C_2, D_2, E_2, F_2, G_2, H_2)$.

The estimated S/N ratio in the optimal condition is 18.52 dB, while that in the initial condition is 16.39 dB. Thus, the gain in S/N ratio between optimal and initial conditions is 2.13 dB, which means that variation in the optimal condition is 20% less than that of the initial condition. This improvement was acceptable but, in fact, not terrific. The next step was to

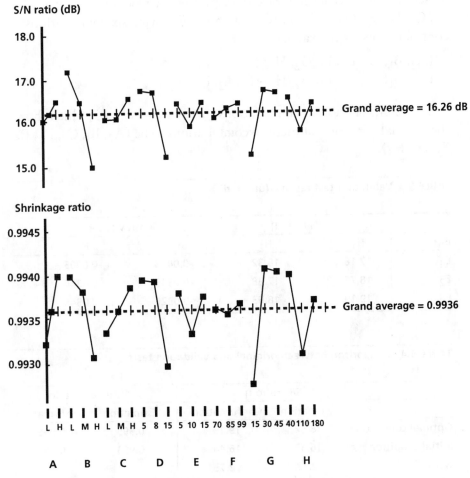

Figure 9.7 Main-effect plots of S/N and shrinkage ratios.

conduct a validation test. We injected plastic test pieces under both optimal and initial conditions, measured the dimensions of the injection mold and injected test pieces, and then calculated S/N ratios and sensitivities for these test pieces. We expected to reproduce the V-shape plots of S/N ratios for both factors E and H. If such were the case, we could trust the result. In other words, the optimal condition would be very stable and robust against noise factors. Validation tests were conducted as follows: Factor H was cooling time and was expected to be reduced to the minimum so as to enhance the productivity of the injection-molding technology. Thus, we selected only levels 1 and 2 for factor H in the validation test. In addition to factor C, we selected all three levels for factor E. All the other factors were set at $(A_2, B_1, C_2, D_1, F_3, G_2)$. We ended up with the following six combinations of control factors for the validation test:

$(A_2, B_1, C_2, D_1, F_3, G_2, H_1)$ $(E_1, E_2, E_3)$
$(A_2, B_1, C_2, D_1, F_3, G_2, H_2)$ $(E_1, E_2, E_3)$

The results of these six combinations are listed in Table 9.3, which shows the optimal condition of these sixcombinations to be $(A_2, B_1, C_2, D_1, E_3, F_3, G_2, H_1)$.

**TABLE 9.3  Validation-test results  (unit = dB)**

| E\H | S/N ratio $\eta$ | | Sensitivity S | |
|---|---|---|---|---|
| | $H_1$ | $H_2$ | $H_1$ | $H_2$ |
| $E_1$ | 17.16 | 18.32 | -0.0478 | -0.0305 |
| $E_2$ | 18.76 | 17.88 | -0.0345 | -0.0340 |
| $E_3$ | 19.19 | 18.63 | -0.0320 | -0.0346 |

**TABLE 9.4  Comparison between original and validation tests**

| | S/N ratio $\eta$ | | Shrinkage ratio | |
|---|---|---|---|---|
| | Estimation | Confirmation | Estimation | Confirmation |
| Optimal condition | 18.52 | 19.19 | 0.995 | 0.996 |
| Initial condition | 16.39 | 16.44 | 0.993 | 0.994 |
| Gain | 2.13 | 2.75 | | |

The validation tests confirmed the gain in S/N ratio between optimal and initial conditions, as illustrated in Table 9.4. However, the project did not end here. We still needed to confirm that the dimensional accuracy of this technology really satisfied the accuracy requirements of potential future products.

When we injected certain test pieces, we found that their shrinkage ratios and dimensions showed significant variation between different directions and locations. This meant that this technology was still not sufficiently robust for mass production, so we tried to identify a better combination of control factors to reduce shrinkage variation.

It is very time inefficient to separately adjust mold dimensions at various locations or directions to compensate for shrinkage variation. To simplify the experiment, we compounded different locations and directions into a single noise factor. We called this compound noise factor location/direction combination, and used the following three noise factors to simulate shrinkage variation in the injection-molding technology:

- Dimensions along $x$ axis at bottom section of the test piece
- Dimensions along $x$ axis at top section
- Dimensions along $y$ axis

**TABLE 9.5    S/N and shrinkage ratios of different location/direction combinations**

|  | S/N ratio (dB) | Shrinkage ratio |
|---|---|---|
| $X$-direction at bottom section | 41.31 | 0.999 |
| $X$-direction at top section | 38.83 | 0.995 |
| $y$-direction (height of test piece) | 31.01 | 0.995 |

Table 9.5 summarizes S/N ratios and shrinkage data from the tests. From this table, we can see that there is a difference among the shrinkage ratios of the three noise factor levels. Because of this shrinkage variation, we were not able to further increase the S/N ratio in the optimal condition of Table 9.4. In other words, the shrinkage variation among different locations and directions reduced the functional robustness of this CFRP injection-molding technology. We had already applied control factors to maximize the functional robustness of this injection-molding technology, and therefore, after this stage, we could only measure the shrinkage ratios of different loca-

tions and directions and use them to adjust mold dimensions to further improve the accuracy of the injection-molding technology.

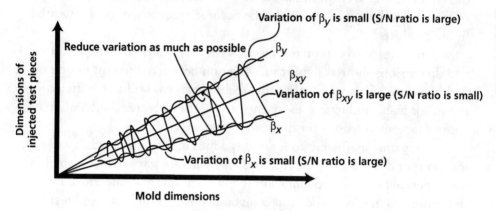

**Figure 9.8** Shrinkage-ratio variation among different directions.

In most injection-molding applications, engineers and mold makers assumed shrinkage ratios of different locations and directions to be the same. However, from Table 9.5 and Fig. 9.8, we can see that the shrinkage ratios along the x axis and y axis of test pieces are different. This shrinkage varia-tion precluded the possibility of increasing accuracy of the injection-mold-ing technology. The optimal condition of injection-molding technology is the one that minimizes shrinkage variation among different directions and locations. If we do not conduct any robust engineering activities, the shrink-age ratios of different directions or locations will differ significantly. The measurement of shrinkage ratio variation in Fig. 9.8 is illustrated in detail in Chapter 4 of *Transformability and Technology Development*, published by the Japanese Standards Association. The results of shrinkage variation among different direction/location combinations are summarized in Table 9.6.

From Table 9.6, we can see that there is significant variation among the shrinkage ratios of different location/direction combinations and that this variation affects the S/N ratio values. However, the shrinkage variation of the optimal condition is much less than that of the initial condition; therefore, the S/N ratio of the optimal condition is larger (better) than that of the initial condition. From these two S/N ratios, we can calculate the accuracy of the injection-molding technology at initial and optimal condi-tions. Initial-condition accuracy is about 40 micrometers, while optimal-condition accuracy is about 20 micrometers. We measured the dimensions of the top and bottom sections of test pieces and confirmed the accuracy estimation. In fact, the accuracy of product dimensions along the y direction was even better, with an accuracy of about 10 micrometers.

**TABLE 9.6  S/N and shrinkage ratios among different location/direction combinations**

| | S/N ratio (dB) | Shrinkage ratio | | |
| --- | --- | --- | --- | --- |
| | | X direction at bottom section | X direction at top section | Y direction |
| Optimal condition | 33.58 | 0.999 | 0.995 | 0.995 |
| Initial condition | 28.19 | 0.999 | 0.993 | 0.993 |
| Gain | 5.39 | | | |

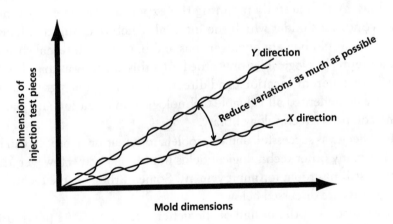

**Figure 9.9** Minimizing shrinkage variation along different directions.

At the technology-development stage, we measured the basic function (i.e., transformability) of the injection-molding technology and tried to determine optimal conditions so as to minimize shrinkage variation (Fig. 9.9). If the shrinkage ratio shows little variation, the internal structure of the injected products should be very homogeneous and of very consistent strength. Strength is a downstream quality characteristic and should be measured at more downstream development stages. After injecting several CFRP test pieces, we conducted destructive tests and measured their strength to determine if they met design standards. We also took cross sections of the injected test pieces from optimal and initial injection-molding conditions and examined them under a microscope. We found that the internal structure of the test pieces formed under optimal conditions was much more homogeneous than under initial conditions. The cross sections of the optimal condition pieces were very delicate and had no internal shrinkage. As a result, we met the downstream quality requirements indirectly through upstream quality activities (i.e., improving the functional robustness of this injection-molding technology).

Before this robust engineering project, the plastic engineers had already used trial-and-error methods to try to solve the problems of internal shrinkage. However, they were not completely successful and the problems still occurred occasionally. In this robust engineering project, these engineers and the author used the concept of transformability and robust engineering methods to improve the functional robustness of this injection-molding technology. After carefully planning the experiments, we determined optimal conditions under which the internal structure of injected CFRP components would be very homogeneous and have little internal shrinkage. Another technological lesson learned from this project was that shrinkage ratios of different locations and directions might differ significantly. Finally, we documented all valuable technological information and placed it in the company's technology library.

This project was successful, and its problem formulation approach can be applied to many other technological fields in a similar way. However, this case study still has room for improvement. Some of the possible technical improvements are discussed below.

First we discussed the design of the generic test piece. As previously mentioned, we needed to measure both the dimensions of injection molds and the corresponding dimensions of injected CFRP test pieces. The dimensions of injection molds are input signals and the dimensions of the injected products are the output responses. We expected the levels of input signals and output responses to vary widely. For easy and accurate measurement of the dimensions of injection-molds and injected test pieces, the test pieces consist of simple geometrical shapes, such as triangles. Of course, several triangles combined can form complicated shapes, such as hexagons or pentagons, as illustrated in Fig. 9.10. The bottom section of the generic test piece in Fig. 9.4 is a pentagon and is not symmetrical along its center line. In general, it is more difficult to maintain the dimensional accuracy of a nonsymmetrical shape than that of a symmetrical shape. In other words, the dimensional accuracy of a symmetrical shape is always better.

After this project, the author concluded that the nonsymmetric pentagon may be superior to the pentagon (ABCDE) or to the two hexagons (8 and 12 mm in thickness) in Fig. 9.4 because we can get more dimensional levels. For example, the pentagon in Fig. 9.10 has 7 levels (1 = 3, 4, ... 9) while the pentagon in Fig. 9.4 has only 3 levels (level 1: AB = AE = ... = CD; level 2: BE = BC = CE = AD; level 3: BD = AC). In the pentagon in Fig. 9.4, the biggest dimension (e.g., BD) is only twice the smallest (e.g., AB). Thus, input signal range is not very wide. In comparison, in Fig. 9.10, the biggest dimension (1 = 9) is three times the smallest (1 = 3). A real product may not, in fact, be composed of symmetrical shapes. Thus, it is better to use nonsymmetrical shapes to develop generic test pieces for the development of robust technologies. Figure 9.10 shows a good, nonsymmetrical shape for the development of injection-molding technology or other technologies.

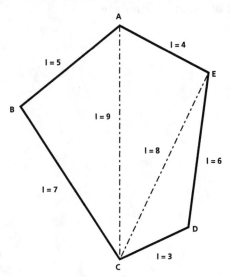

**Figure 9.10** Proposed shape for a generic test piece for injection-molding technology. (Note: Shape is composed of several triangles.)

Another issue is related to calculation of the S/N ratio (measurement of robustness) for shrinkage variation of the test pieces. Some engineers thought that we should calculate one S/N ratio for each direction or location, but this was not an appropriate robust engineering approach. The S/N ratio is a measurement of the overall functional robustness of the whole injection-molding technology, so it should include all possible noise conditions, such as shrinkage variation in all directions and locations. Ideally, we expected all shrinkage ratios in all directions and locations to be the same. Thus, we should use only one S/N ratio to measure the functional robustness of the whole injection-molding technology. If we can maximize the S/N ratio, we will be able to simultaneously reduce the shrinkage-ratio variation of different directions and locations. If we apply several S/N ratios to separately measure the shrinkage variations of various directions or locations, we will be unable to determine as many control factors to simultaneously reduce these shrinkage variations. In a real-life product-development process, it is impossible to equalize shrinkage ratios of all locations or directions if the basic function of the injection-molding technology is not robust. To sum up, we should use only one S/N ratio to measure the overall shrinkage variation of each test piece. This is a very important point in the development of robust technology.

**CHAPTER**

# 10

# The Future
# of
# Robust Engineering

The use of robust engineering to develop reliable and generic technologies will undoubtedly be a major trend in manufacturing industries all over the world. The manufacturing industries of the United States suffered from low quality and low productivity during the 1980s in their competition with Japanese industries. However, those manufacturing companies are now devoting themselves to the enhancement of productivity through the development of new technologies and organizational restructuring. It is more cost and time efficient to develop robust technologies through cooperation among different manufacturing companies than individually. Currently, American manufacturing industries are working together (even with competitors) to develop basic manufacturing technologies. Any manufacturing company that does not have a base of robust technologies cannot compete in the highly competitive markets of the next century. As a result, an International Productivity Conference is now held every five years to enhance cooperation among manufacturing industries through the application of robust engineering for technology development.

Currently, Japanese manufacturing industries are still ahead of their American competitors in terms of quality-engineering capability. However, the technological capability of American manufacturing industries is much more advanced than in Japan because of the tremendous investment of human and engineering resources in basic research. If their technological capability can be efficiently applied to develop commercial products, they will be most formidable. The productivity of American industries is based on basic scientific research and development, not on the development of specific products. Because most of their engineering resources are not focused on the development of specific products, they do not pay as much attention as their Japanese competitors to the time-to-market span or the reproducibility (i.e., robustness) of their current products. Currently, the average quality level of American products is still not as high as that of Japanese products, but if the manufacturing industries of the United States can shift their quality-engineering paradigms upstream and apply robust engineering to develop robust and flexible technologies, they will be able to achieve both high quality and low cost simultaneously. As a result, they will become competitive in the global market for commercial products.

By comparison, the manufacturing industries of Japan are still based on product-oriented development (though new technologies are now being developed at a fast rate). Because their major resources are focused on the development of specific products, they are more concerned with the time-to-market span and the quality of specific products rather than with the development of generic technologies. In other words, the success of Japanese manufacturing industries is based on the exhaustive efforts and constant overtime (as illustrated in Fig. 10.1) of Japanese engineers and managers in the development and continuous improvement of specific products. To maintain international competitiveness, Japanese manufacturing industries need to develop a new generation of robust technologies for future products. To achieve this, robust engineering needs to be emphasized at the development stage of new technologies. One important future task for Japanese manufacturing industries is to shift robust engineering from product-oriented development to technology-oriented development. Only through this approach can engineers improve the efficiency of their total-product-development processes.

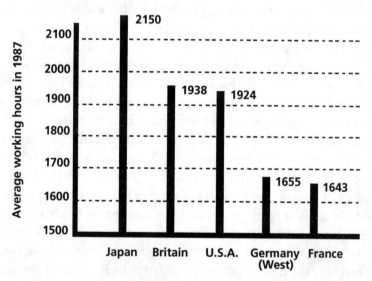

**Figure 10.1** Average working hours of the major industrialized countries. (*Modern News Weekly,* December 23, 1989, Koudansha Co., Tokyo, Japan.)

One potential problem for manufacturing industries in the next century is the aging work force (Fig. 10.2). The average age of the work force in the major industrial countries is increasing rapidly because of advances in health and medical care and lower birth rates.

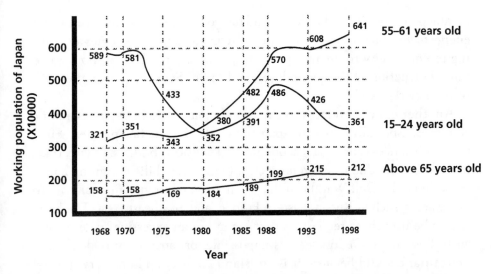

**Figure 10.2** Working-population distribution (Japan). (From "The Salaried Employee of the Twenty-First Century," Economic Planning Agency, Planning Bureau, Tokyo, Japan, 1992.)

The aging problem may mean that the average worker in the manufacturing industries of the next century may not be able to work as long daily. For productivity, it is critical to develop product and process technologies that are easy for this aging work force to use, so that they can develop and deliver products efficiently. If a base of robust product and process technologies is developed, engineers and other workers will be able to develop and manufacture high-quality products in a time- and cost-efficient way. Developing robust technologies is the most efficient way to ensure the productivity of the manufacturing industries of the next century.

Other future tasks facing robust engineering lie in the development of new-energy technologies, such as the development of new-energy resources and prevention of environmental pollution. Current energy technologies are primarily built around internal combustion engines and petroleum energy. Petroleum energy is widely applied to generate electric power and to provide power to airplanes, ships, and most transportation vehicles. It is certain that the earth's petroleum will be used up sooner or later, and new-energy technologies must be developed to replace petroleum-energy technologies. One potential way to generate high-quality energy without negative side effects is solar energy. Solar energy will cause little environmental pollution, is inexhaustible, and has good potential to replace current energy technologies. Currently, internal-combustion technologies are causing too many environmental problems, such as air pollution, noise, acid rain, destruction of the ozone layer, and so on. Sooner or later, they will have to be replaced.

Another concern relates to the tremendous trade imbalances among countries, especially between Japan and many of its trading partners. Such trade imbalances cause many economic and political conflicts. In fact, trade imbalances are due to unequal technological capabilities among countries. Since the Gulf War, Japan has been asked to provide financial aid to many areas, such as Russia, Eastern Europe, and Southeast Asia. The basic reason behind these requests is that Japan has an excessive trade surplus with these areas. Of course, it is impossible for Japan to provide permanent financial aid because it is an island country and has very limited resources. To fundamentally solve these trade imbalances, we must help these countries to develop their own manufacturing-technology bases and thereby their manufacturing industries. In other words, the industrialized countries need to work together to help developing countries build their own bases of manufacturing technologies. This fundamental step toward balancing trade among countries will reduce possible economic and political conflicts.

In this way, the developing countries will have the opportunity to develop their industries and economies, living standards can be raised to the level of the industrialized countries, and we can but hope that conflicts will decrease. But these goals depend on developing countries having competitive manufacturing industries, as commercial technology capability is critical to economic power. Technological competitiveness of a manufacturing company depends on two major criteria: manufacturing cost and quality. Robust engineering is the most efficient way to improve the development efficiency of new technologies, by which manufacturing cost can be reduced as the quality of future products is improved. A little technological know-how does not make a manufacturing company technologically competitive. To compete in the next century, a manufacturing company needs to develop a profound technology base, composed of flexible and robust technologies that are ready and in place to assist in the development and manufacturing of future products. This is the theme of robust engineering for technology development.

To improve manufacturing productivity, we need to develop well-planned technologies that can be easily used by product engineers for the development of future products. If robust technologies are developed, the efficiency of product-development processes will improve significantly. Currently, the Ministry of International Trade and Industry of Japan is leading a giant project, the Intelligent Manufacturing System (IMS), which the United States, Canada, Australia, and some European countries have been invited to join. Its objective is to build artificial intelligence into manufacturing technologies and thereby develop a new generation of manufacturing systems. Robust engineering will play a very important role in this giant project.

As mentioned in previous chapters, technologies are developing faster than ever now. Training engineers quickly in the development of robust technologies and smoothing out the interface between technology development and product development are two critical issues in robust-technology development, and IDS (the Intelligent Design System) may be the solution. The capability of most engineers to design and develop new products is usually proportional to their industrial experience. However, if new technologies can be made robust at a very early development stage, they will be flexible and will enhance engineers' development and design capabilities; this means that engineers, regardless of industrial experience, will be able to use such robust technologies to rapidly develop new products. Technology education is another important issue. We will need to use many educational methods to transfer knowledge from one engineering generation to the next, to ensure efficient communication of technology information.

New technologies should be highly flexible to allow for the efficient development of new products, and robust engineering is the most efficient way to enhance this flexibility. Therefore, all engineers (from advanced research engineers to production-process engineers) must be familiar with robust engineering for use in product development, product design, and manufacturing-technology development.

In conclusion, robust engineering should be the common tool for the development of all product and manufacturing technologies of manufacturing companies everywhere.

# INDEX

## A

Accurate measurements, robust technology development and, 148–149

Air pollution, 18–21

Automotive brake pads, 117–121

Automotive industry, 3, 16–18, 117–121; *see also* Manufacturing industries; Reliability engineering

Kenzo Ueno's background in, 3–11

robust technology implementation, 137–139, 144–160

society and quality problems; *see* Society, quality and

## B

Ball joints, 113–116

Basic functions of a technology; *see* functionality

## C

Case studies; *see* individual topics

CFRP injection-molding, 144–160

Challenging projects, 79–80, 83–85

Classifications, of quality; *see* Quality

Company-wide implementation, 75–90

case-study workshops for, 80–81, 85

challenging projects for, 79–80, 83–85

consulting on, 85–87

impact of, 87–90

introduction phase, 78–83

popularization of, 81–83, 85–87

projects and, 79–80, 83–85

resistance to, 86–88

selective training and, 83–84

upstream problem solutions and, 85

Competitiveness, 7, 45, 140–143, 166–167

Confirmation tests, 86, 100, 101–109; *see also* Validation tests

Control factors, 84–85, 128–133

optimal settings for, 67–68, 86

output response nonlinearity and, 68–73, 128–131, 146–147, 150–157

Cost/quality relationship, 99–100

Customer satisfaction, 7, 16–18, 22, 37–38, 97–99

## D

Design parameters; *see* Control factors; Robust engineering

Deterioration variation; *see also* Noise
(disturbance) factors
performance deterioration,
7–8, 42–43
Developing countries, 166
Development; *see* Product
development
Diagnosis of problems, prevention vs.,
7–8
Disturbance factors; *see* Noise
(disturbance) factors
Downstream issues, 3, 32–34
parameter design and, 66–67
product development shift, 64–66
quality
characteristics, 33–34, 97, 125
upstream quality vs., 85, 94–95
technology-development, 33–34
Durability, 5, 7–8, 16–18;
*see also* Reliability
Dynamic robust engineering,
121–133, 142

**E**

Education, engineering; *see*
Engineering Efficiency,
in product development, 44, 59
Electronic circuits, 29–30
Electronic control devices, 137–138
Electronic weapons, 15–16
Electronification, 21–22
Emission control, 18–21
Energy technologies, 165–166
Engineering, 7–10, 39–41
downstream, 32–34
education, 9, 10–11, 42, 44
Environmental concerns, 8, 18–21,
42–43
Experimental designs, 70–71

**F**

Failure modes and effects analysis
(FMEA), 15–16
Fire-fighting, noise compensation
and, 43
Flexibility, robust technology and,
98–99, 167
Friction force, 113–116
Front-suspension technology, 137
Fuji-Xerox, 41
Functionality, 32, 63–64, 99
basic, of a technology, 111–133
automotive brake pads, 117–121
ball joints, 113–116
case study, 121–133
control factors and, 128–133
enhancement, 113–121
machining, 115–116
NC machines, 126–128
proportionality constant ($\beta$)
and, 127–128
quality problems and, 113
robustness enhancement
examples, 113–121
loss of, 93
measuring, 57–58
quality and, 96
variations in, 8–9, 71–72,
106–109, 138–139

**G**

Generic technologies, 125, 163
Generic test pieces, design of,
158–159

# I

Ideal functions, of new technology,
55–59
Imai, Kaneichiro, 10
Implementation, company-wide; *see*
Company-wide implementation
Industrial education, 10–11
Initial conditions, 7–8, 32, 129–131
Injection-molding case study,
144–160
Input; *see* Control factors
Integrated circuits (ICs), 29–30
Isobe, Kunio, 41–42

# J

Japanese Administration and
Management Research Institute,
41–42
Japanese Industrial Education
Association, 10

# L

Laser-welding case study, 45–59
Lead time, 32–33
Life-cycle tests case study, 101–109

# M

Machining, 115–116, 121–133
Management resistance, to robust
engineering, 86–87
Manufacturing industries, 38,
139–143, 163–167;
*see also* Automotive industry
Mass-produced products, 4, 5, 28–29,
103–104, 140
Mean-value methods, 40–41, 48–49
Measurements, 8, 148–149

Middle-level management resistance,
to robust engineering, 86–87
Midstream quality characteristics, 97
Motorization, 16–18
Murayama Proving Ground, 4–7

# N

New-product designs, 10
New technologies, 27–33, 37–39,
165–166; *see also* Technology
customer satisfaction and, 37–38
ideal functions of, 55–59
product development and, 30–33
reliability engineering and,
23–24, 27–28
Nissan Motor Co., 3, 9, 77, 82–83,
87–90
Noise (disturbance) factors, 8,
33–34, 42–43; *see also* S/N ratios
product desensitization against, 37,
151–154
robustness against, 63–64
two-level, 52–54, 128–129
Nonlinearity effects, output, 68–73,
128–131, 146–147, 150–157
Numerical control (NC) machining,
121–133

# O

Optimization, of control factors,
67–68, 86, 129–131
Origin quality characteristics, 96–97
Orthogonal arrays, 86, 128–129
Output responses, nonlinearity of,
68–73, 128–131, 146–147, 150–157
Overseas plants, 38

# P

Paradigms, shifting, 9
Parameter design; *see* Control factors;
   Robust engineering
Performance, 7–8, 17–18, 42–43
Plastics, injection-molding, 144–160
Popularization, of robust engineering,
   81–83, 85–87
Problem diagnosis, vs. problem
   prevention, 7–8
Product desensitization, against noise,
   37, 151–154
Product development, 43, 44, 59
   downstream quality and, 33–34
   new technology and, 30–33
   robust technology method, 49–52
Product failure prevention,
   100–101
Production, transferred overseas, 38
Production-oriented development,
   vs. technology-oriented, 64–66
Productivity, 44, 59, 167
Product planning, 143
Projects, for robust engineering,
   79–80, 83–85
Proportionality, 107–108, 126
   constant (β), as basic function
   descriptor, 127–128
Prototypes, 4, 5, 102–103
Proving grounds, 3–5

# Q

Quality
   assurance, 4, 7–8, 39, 43
   classifications and requirements,
      91–109
      case study, 101–109

cost/quality relationship, 99–100
customer satisfaction and, 97,
   98–99
definition of quality, 93
downstream quality; *see*
   Downstream issues
flexibility and, 98–99
functionality, 96; *see also*
   Functionality
midstream quality, 97
origin quality, 96–97
product failure prevention,
   100–101
technological readiness and, 98
upstream quality, 97
initial, performance deterioration
   vs., 7–8
problems, 113
   action strategies and, 35–59
   case study, 45–59
   new technology and, 37–39
   noise factors and, 37; *see also*
      Noise (disturbance) factors
   science and engineering, 39–41
   technology bases, changing,
      43–45
   variation and, 41–43
reliability; *see* Reliability
and society; *see* Society,
   quality and
Quality-by-inspection, 32
Quality engineering; *see* Robust
   engineering

# R

Reeducation, of engineers, 44
Reliability, 5–8, 16–18, 27–28

Reliability engineering; *see also*
  Reliability
  automotive industry, 13–24
  new technologies and, 23–24,
    27–28
  WWII electronic weapons and,
    15–16
Repeatability, 99; *see also*
  Functionality; Robustness
Requirements, quality; *see* Quality
Restructuring of technology bases;
  *see* Technology bases
Robust engineering, 8–9, 72; *see also*
  Robust technology development
    advantages of, 73
  company-wide implementation;
    *see* Company-wide
    implementation competitiveness
    and, 7, 45, 140–143, 166–167
  downstream, 32–33
  dynamic, 142
  future of, 161–167
  for new technologies, 39
  nonlinearity effects, output vs.
    control factors, 68–73,
    128–131, 146–147, 150–157
  parameter design for, 61–74
  product failure prevention,
    100–101
  reasons for, 9–11
  technology base restructuring and;
    *see* Technology bases
Robustness; *see also* Functionality;
  Repeatability confirmation tests;
  *see* Confirmation tests
  enhancement of, experimental
  design for, 70–71

Robust technology development,
  49–52; *see also* Robust engineering
  accurate measurements and,
    148–149
  factors critical to, 98–99
  ideal functions for new technology,
    55–59
  implementations of, 135–160
    automobile industry, 137–139
    case study, 144–160
    manufacturing industries,
      139–143, 163–165

**S**
Science, quality problem solutions
  and, 39–41
Selective training, 83–84
Sensitivity, 128–129
S/N ratios, 56, 129–133, 151–157,
  160; *see also* Noise (disturbance)
  factors
Society, quality and, 25–34
Stability, 32; *see also* Functionality
Standards, testing, 3–5
Static welding strength case study,
  45–59
Statistical analysis methods,
  reliability improvement and, 6
Statistical process control, 32

**T**
Taguchi, Genichi, 8–9, 64, 80–81, 88,
  121, 131–133
Technological readiness, 98
Technology, 23–24; *see also*
  New technologies,
  basic function of; *see* Functionality

Technology bases, 3–5, 8–9
    emptying, 43–45
    reliability improvement and, 6–8
    restructuring, 1–11, 43–45, 51
Technology-development engineers,
    33–34
Technology-oriented development,
    production-oriented vs., 64–66
Tenth Taguchi Symposium, 121,
    131–133
Testing departments, 3–5, 7–8
Top-down popularization of robust
    engineering, 85–87
Total quality loss, 93
Toyota Motor Co., 77
Trade imbalances, 166
Training, in robust engineering,
    83–85, 167
Tsuchiya, Mr., 41

## U
Ueno, Kenzo, automotive industry
    background, 3–11
Upstream issues, 85, 94–95, 97
Usage loss, 93

## V
Validation tests, 57, 59, 101,
    103–104; see also Confirmation tests
Variation, 41–43; see also
    Functionality

## W
Welding, 45–59
Workshops, 80–81, 85
WWII electronic weapons, 15–16